GENE EXPRESSION
and ITS CONTROL

SERIES EDITORS
Kay E. Davies, *University of Oxford*
Shirley M. Tilghman, *Princeton University*

The identification and mapping of genes, analysis of their structures, and discovery of the functions they encode are now cornerstones of experimental biology, health research, and biotechnology. *Genome Anaysis* is a series of short, single-theme books that review the data, methods, and ideas emerging from the study of genetic information in humans and other species. Each volume contains invited papers that are timely, informative, and concise. These books are an information source for junior and senior investigators in all branches of biomedicine interested in this new and fruitful field of research.

SERIES VOLUMES
1. Genetic and Physical Mapping
2. Gene Expression and Its Control

Forthcoming

3. Genes and Phenotypes
4. Strategies for Physical Mapping

GENE EXPRESSION and ITS CONTROL

Edited by
Kay E. Davies
University of Oxford

Shirley M. Tilghman
Princeton University

Volume 2 / GENOME ANALYSIS

 Cold Spring Harbor Laboratory Press 1991

Genome Analysis Volume 2
Gene Expression and Its Control

All rights reserved
Copyright 1991 by Cold Spring Harbor Laboratory Press
Printed in the United States of America
ISBN 0-87969-359-2
ISSN 1050-8430

Cover and book design by Leon Bolognesi & Associates, Inc.

Authorization to photocopy items for internal or personal use, or the internal or personal use of specific clients, is granted by Cold Spring Harbor Laboratory Press for libraries and other users registered with the Copyright Clearance Center (CCC) Transactional Reporting Service, provided that the base fee of $3.00 per article is paid directly to CCC, 27 Congress St., Salem, MA 01970. [0-87969-359-2/91 $3.00 + .00]. This consent does not extend to other kinds of copying, such as copying for general distribution, for advertising or promotional purposes, for creating new collective works, or for resale.

All Cold Spring Harbor Laboratory Press publications may be ordered directly from Cold Spring Harbor Laboratory Press, 10 Skyline Drive, Plainview, New York 11803. Phone: 1-800-843-4388. In New York (516) 349-1930. FAX: (516) 349-1946.

Contents

Preface *vii*

Embryonic Stem Cells as a Route to Experimental Mammalian Genetics   1
Martin Evans

Insertional Mutagenesis   13
Michael A. Rudnicki and Rudolf Jaenisch

Chromosome Imprinting and Its Significance for Mammalian Development   41
Bruce M. Cattanach

The X-inactivation Center and Mapping of the Mouse X Chromosome   73
Stephen D.M. Brown

The Regulation of the Human β-globin Locus   99
Niall Dillon, Dale Talbot, Sjaak Philipsen, Olivia Hanscombe, Peter Fraser, Sara Pruzina, Mike Lindenbaum, and Frank Grosveld

Index *119*

Preface

Intensive efforts by many investigators are currently being applied to generate high-resolution genetic and physical maps of two mammalian species, human and mouse. These maps can be considered a compilation of road marks, whose usefulness depends on the average distance between the marks. At the highest level of resolution, they provide immediate molecular access to any region of the genome. Thus, one can realistically anticipate the day when the assignment of a gene or a trait to a chromosomal region will lead to its rapid cloning. Yet this will be a hollow victory if we do not develop in-parallel technologies for understanding the mammalian genome. The commitment of genome organizations throughout the world to the parallel analysis of the human and mouse genomes reflects an appreciation of this fact; i.e., the isolation of a human gene is just the beginning of a much longer process of understanding the function of the gene within the context of the organism. By general consensus, the mouse has become the model mammalian organism in which experimental tests of the function of specific genes will usually be conducted.

This volume is intended to provide a glimpse into the future for a look at how functional analyses of genes and genomic regions are likely to proceed and to emphasize the power of the comparative approach. The chapter by Frank Grosveld and his colleagues illustrates that patterns of transcription of human genes can be studied by transposing the genes into the mouse. The mechanisms underlying seemingly more complex phenomena such as genomic imprinting and X chromosome inactivation are beginning to be amenable to functional analysis as well, as illustrated in the chapters by Bruce Cattanach and Stephen Brown. Finally, to understand the function of a specific gene completely, it is necessary to generate specific mutations in it. Until recently, this has not been possible in the mouse, nor, in fact, in any multicellular eukaryote. The development of mouse embryonic stem cells which can shuttle between a tissue-culture dish and a mouse blastocyst coupled to the application of homologous recombination for specific gene replacement marks the beginning of a new era in mouse genetic analysis, as described in the chapter by Martin Evans. The identification of mutations in completely new genes through insertional mutagenesis, reviewed by Michael Rudnicki and Rudolf Jaenisch, is likewise certain to yield fertile ground for understanding gene function.

We are grateful to all of the authors in this volume for their hard work and excellent contributions. We are also grateful to the staff of the Cold Spring Harbor Laboratory Press, especially Nancy Ford and her colleagues Dorothy Brown, Virginia Chanda, and Mary Cozza, who have been such enthusiastic partners in the publication of this series, and who have ensured that the series is produced in such a timely fashion.

Kay E. Davies
Shirley M. Tilghman
April 1991

GENE EXPRESSION
and ITS CONTROL

Embryonic Stem Cells as a Route to an Experimental Mammalian Genetics

Martin Evans

The Wellcome Trust, Cancer Research Campaign
Institute of Cancer and Developmental Biology
and Department of Genetics
University of Cambridge
Cambridge CB2 1QR, England

Transgenesis by the use of embryonic stem cells, whereby specific gene alterations may be preselected in vitro, coupled with the more conventional microinjection of DNA into a zygote, now provides a route to a complete experimental genetics system in the mouse. This particular combination of molecular genetics, tissue culture of pluripotential embryonic cells, and whole-animal genetics has opened up a completely new route to exploration of genetic function and dissection of the complex developmental and physiological interactions in the whole animal.

This chapter discusses:

❏ embryonic stem cells from the mouse and other animals

❏ gene targeting in embryonic stem cells and methods for screening and selection of the cell clones carrying the desired genetic alteration

❏ examples of gene loci that have been targeted successfully and use of new mutations for the creation of animal models of genetic disease and for genetic analysis of development

❏ potential application of similar technologies to other species

❏ technical limitations and possible resolutions

INTRODUCTION

Embryonic stem cells

First isolated and described by Evans and Kaufman (1981), embryonic stem (ES) cells are tissue-culture cell lines of totipotential cells derived from an early embryonic epiblast. They maintain their totipotentiality in tissue culture and, upon reinoculation into a suitable embryo, give rise to chimeric mice in which all the tissues, including the germ cells, have contributions from the original ES cells (Bradley et al. 1984). Thus, a complete route is established between tissue culture and the whole-animal genome.

Germ-line transgenesis

To provide a route to genetic transformation of any specific tissue, it is necessary to transform a cell which is the progenitor of that tissue. Thus, to transform the germ-cell line, it is necessary to transform a cell in the germ-cell lineage. Transgenesis in mammals has therefore concentrated either on the early embryo at a stage when its cells precede any determination into somatic or germ-cell lines or upon transformation of a cell in the germ-cell lineage. In the intact embryo, either technique requires transforming cells that are present in small numbers and are inaccessible to experimental manipulation. They are, moreover, present only transiently in the process of development of the embryo. Of those cells that lead to the germ-cell lineage, to date only ES cells have been successfully maintained as tissue-culture cell lines.

A tissue-culture system has two distinct advantages over experimental systems using a normal embryo: (1) availability for unlimited cell numbers and (2) the protracted time over which manipulations and selections may be carried out. It is therefore possible to transform these cells genetically and to select or screen for infrequent events that lead to the production of the desired genetic alteration (Hooper et al. 1987; Kuehn et al. 1987). Coupled to this is the observation that homologous recombination may take place in mammalian cells between an incoming DNA construct and endogenous chromosomal genes (Smithies et al. 1985). This recombination allows the generation of specific gene alterations and, together with the more conventional procedure of zygote microinjection, has created a methodology for a completely experimental genetics system in the mouse. This work is being rapidly extended to other experimental and domestic animals.

Mammalian genetics

Notwithstanding the great theoretical and practical interest in mammalian genetics, the conventional genetic analysis of mammals has been handicapped vis-à-vis other experimental animal systems because of relatively small litter sizes and long life cycles and in the case of human genetics, its nonexperimental nature. Mammalian genetics has, however, benefited from the study of human genetics. Although there is no experimental breeding in human systems, detailed observation of a very large population has allowed investigation both of polymorphisms and of very rare mutations, and statistical methods for analysis of data available from human genetic pedigrees have become highly refined (McKusick 1991).

Traditional genetics depends on mutations or preexisting genetic polymorphisms that are discovered in a species. Animal breeding depends on selection from suitable variation, either preexisting in or deliberately introduced into the stock. The major tools for genetic analysis are therefore breeding segregation studies and direct phenotypic analysis. The only experimental approach to widen the scope of genetic variants available for study is mutagenesis followed by specific screening or fortuitous recovery of relevant alleles. Added to this, methods for nonmeiotic genetic analysis—somatic cell genetics and, more recently, direct molecular biological analysis—have been very effectively applied as well as being combined with pedigree analysis so that mammalian genetic maps and knowledge of gene sequence data are advancing rapidly.

Our understanding of genetic function and the practical application of our genetic knowledge in experimental animals require the ability to modify the genome deliberately, preferably in a manner not entirely reliant upon screening the accidents of nature. Thus, the concept of a reverse mammalian genetics emerges, where the effect of specific genetic modification may be studied in the context of the intact organism. Genes of interest are now being identified from the results of intensifying mammalian genetic analysis, through molecular biology techniques, and by cross-homology with similar genes of other species. Both through the identification of important protein products and through direct gene mapping, many loci have been identified in the normal genome for which there are no known mutations and for which there is no known tested function. The only true test of the function in the whole organism is the genetic test of ablation or alteration.

Large classes of mutations in mammals are also systematically unavailable for study. A completely recessive embryonic lethal mutation is one that identifies a gene locus whose function is essential for normal development but which is not haplo-insufficient. Such entirely recessive embryonic lethal mutations are extremely unlikely to be discovered either in humans or in experimental mammals, and the only recessive embryonic lethals seen will be those that have an associated dominant nonlethal phenotype. One might expect that this class of mutation includes many of the most important control loci involved in the specification of the developmental process and that identification of these loci would be necessary to provide a genetic analysis of development. A complete analysis would need to examine the effect of subtle changes in function at these loci and not merely identify any deletion of function. The technology established with ES cells has provided the opportunity for such direct experimental genetic analysis.

In a great number of cases, genes have been identified in mice, either by analogy with those of other species (e.g., human disease syndromes or *Drosophila* genetics) or from the biochemistry of their protein products, for which there is no mutation of protein structure or lack-of-function allele and thus no rigorous genetic test of function. Further functional studies can only be provided by creating such alleles. In the field of practical application to domestic farm animals, potential alteration of normal physiology may be desirable, and the deletion or modification of the function of controlling genes may be just as important as the overexpression of others.

Although work in this area is proceeding apace, it is still a demanding and time-consuming process, and as yet, a relatively small number of projects have come to fruition. In this chapter, I discuss the new experimental techniques and their applications, some of the results reported to date, and predictions about future directions.

AN EXPERIMENTAL MAMMALIAN GENETICS

ES cells from the mouse

The advent of murine ES cell lines and the subsequent demonstration of their functional totipotency expanded the range of techniques available for gene transfer and gene modification in mammals. In mice, these cells provide an alternative route for transgenesis to that of zygote pronuclear microinjection, and they have the advantage that a large range of in vitro genetic transformation and screening or selective techniques may be applied prior to reconstitution into an animal. In addition, techniques of somatic cell genetics can be used to introduce genetic manipulations into viable and fertile animals. This approach offers unique practical opportunities for experimental genetic manipulation of domestic animals, including targeted mutagenesis by selective inactivation or replacement of endogenous genes, the introduction of clonal alterations to the germ line, and improved control of expression of transgenes through screening the phenotypes of cells in culture.

Gene targeting

Both Smithies et al. (1985) and Thomas et al. (1986) were able to demonstrate that homologous recombination between incoming transfected DNA and chromosomal loci could occur in tissue culture of mammalian cells. This allows specific genetic alteration of endogenous gene loci when used together with the technique of germ-line reconstitution from tissue culture using ES cells and provides an extremely powerful method for genetic manipulation of any chromosomal gene in the mouse.

To maintain the capacity for germ-line transformation, euploidy of the cells at least for the autosomes must not be grossly affected, and the introduction of other random mutations should not be so severe as to prevent germ-line chimerism and introduction of the transformed cell line into a breeding pool. Randomly cointroduced mutations may be segregated away from any desired mutation during meiosis. I have previously reviewed the potential genetic manipulation of mammals and the techniques available (Evans 1989).

General principles Gene targeting involves treating the cells with a DNA construct that is homologous to the chromosomal target DNA but which contains a region of nonhomology to provide an alteration in structure. In this way, the new desired mutation can be introduced and the modified stem cell clone can be screened or detected in culture. In most cases, this means that a positively selectable cassette (e.g., conferring G418 resistance) is included within the DNA construct, allowing selection for all cells that become stably transformed with the construct and leaving the problem as one merely of differentiating cell clones in which the incoming DNA has become integrated in the homologous site from those where it has become integrated at random.

Alternatives to the directly linked, positively selectable cassette are cotransfection and selection (Reid et al. 1991) and high-efficiency nonselective techniques such as cellular microinjection (Zimmer and Gruss 1989). DNA microinjection can result in up to 10% of the injected cells becoming stably transformed. Similarly, multiple reinfection with retroviral vector suspensions may result in all or the majority of the cells in a population becoming transfected (Robertson et al. 1986). Direct screening for the desired integration is clearly feasible with these two methods. Other methods, however, lead to transformation of very few cells in the population. Although widely different frequencies of transformation are reported, in our experience, electroporation rarely provides stable transformation of better than 1 in 10^4 ES cells treated, whereas the calcium phosphate method provides approximately one tenth of this rate. Lipofection, although potentially a very efficient method of DNA delivery, has in our hands only given about 1 in 10^4 stable transformants (M. Evans, unpubl.). Because of these low transformation efficiencies, it is necessary to use a dominantly selectable marker to isolate transformed cell clones. The most frequently used marker is the G418 resistance conferred by expression of the neo^r gene originally derived from the bacterial transposon Tn5. Selection for hypoxanthine phosphoribosyltransferase (HPRT) function in HAT medium (hypoxanthine, aminopterin, thymidine) and selection for lack of HPRT function by the use of 6-thioguanine have also been demonstrated to be efficient and nondamaging selective techniques for use with ES cells. It is becoming clear that there is a need for the development of other suitable selection techniques using other dominant markers.

Isolating the targeted clone After genetic addition or targeted alteration is made to the cells in vitro, the required mutated cell clone must be isolated either by screening or by selection prior to reimplantation into the host blastocyst to produce chimeric animals. Although phenotypic screening may be possible in some specific cases, the final analysis must be at the DNA level because the final desired alteration of the cells reflects a specific alteration of the chromosomal DNA.

One of the most sensitive techniques for detection of such alterations is the polymerase chain reaction (PCR) (Saiki et al. 1985). As a screening method, PCR can be used to identify the desired alteration in multiclonal samples and can serve in sib selection techniques for selecting the desired cell clone (Kim and Smithies 1988; Joyner et al. 1989; Zimmer and Gruss 1989). In practice, there are associated problems with PCR both of oversensitivity giving rise to false positive reactions and of the difficulty of being certain of picking up the desired reaction in every case and in every sample. The technique always needs to be followed by verification by, for example, using Southern blotting.

Screening for the desired gene targeting in most cases depends on a PCR fragment of considerable length because this includes one side of the homology in the targeting construct. As the sensitivity and reliability of the PCR are to some extent inversely dependent on the size of the fragment being amplified, this is not an ideal situation. It may be desirable to design the targeting construct with PCR screening in mind and with a specific small indicator mutation near one of the ends of the construct.

Targeted mutagenesis of endogenous genes The first locus to which gene targeting was applied using the ES cell system which resulted in new alleles in the mouse gene pool was the *HPRT* locus. Null alleles at this locus had already been isolated in ES cells in vitro and reduced to the murine germ line, but the advantages of its being selectable in both the forward and reverse directions made this locus a target of choice for the development of gene targeting strategies by homologous recombination. Indeed, the first successful germ-line transmission of a locus introduced into ES cells in vitro by gene targeting was the correction of an

HPRT deletion mutation (Doetschman et al. 1987). Subsequently, gene targeting in ES cells has been used to create a number of mutations, some of which have now been reduced to the breeding germ line. Thus, the action of the mutated gene in vivo has been able to be explored.

A field of particular interest has been the developmental effects and consequences of mutation of various loci to a null phenotype. These have fallen into three natural groups: cellular oncogenes, immune-system-related loci, and loci with a developmental role. For example, cellular oncogenes that have been identified by the effects of their dominant mutations have in most cases no known null mutation, and therefore the normal functional effect of the cellular oncogene is difficult to assess. Because of the drastic effect of mutation and misexpression, it might be thought that such loci represent considerable hostages to fortune and that their maintenance and evolutionary conservation mean that they have essential, irreplaceable normal functions. Numbers of oncogenes have been targeted to null mutations in ES cells to provide a genetic test of their function. The phenotype of the homozygous null mutation has been reported for c-*src*, *Wnt*-1, c-*myb*, and N-*myc* proto-oncogenes (Charron et al. 1990; McMahon and Bradley 1990; Stanton et al. 1990; Thomas and Capecchi 1990; Mucenski et al. 1991; Soriano et al. 1991). c-*src* and *Wnt*-1 have been bred to homozygosity, and in both cases, live mice were born, demonstrating that all the complex processes of fetal cellular development are not dependent on expression of these genes. In the case of c-*src*, the surprising phenotype was that of failure of bone remodeling (osteopetrosis) and with *Wnt*-1, the abnormalities, although severe, were restricted to the midbrain and cerebellum. Both are areas in which the wild-type allele is expressed, but no simple correlation exists between normal gene expression and developmental perturbation.

Mucenski et al. (1991) have shown that c-*myb* null mutation mice develop normally to day 13, but become severely anemic as the hepatic hematogenesis falls (about day 15). That they develop normally this far is surprising in view of the widespread expression of c-*myb* (including ES cells; Dyson et al. 1989) and its postulated role in cell cycle control. These studies show that its critical function is in hepatic, but not yolk sac, hematopoiesis.

Genes involved in immune signaling and response have also provided a fertile field for investigation. Two examples of loci where gene targeting has been used to give experimental novel mutations are the β_2-microglobulin and immunoglobulin µ chain. β_2-microglobulin is an obligatory associate for expression of the major histocompatibility complex (MHC) class I genes on the cell surface, and loss of β_2-microglobulin expression might have been expected to cause severe problems of cell-cell recognition during development and immune system dysfunction. Results from two independent groups in fact show that mice with no β_2-microglobulin develop normally, except for a deficit in cytotoxic T cells (Koller et al. 1990; Zilstra et al. 1990). Kitamura et al. (1991) introduced a disruption of the membrane exon of the immunoglobulin μ chain that, by preventing cell surface expression of IgM on pre-B cells as had been predicted, prevented the establishment of the B-cell arm of the immune system in the mice.

A third particularly active area of gene targeting is aimed at testing the function of loci with a putative developmental role, such as the homeobox loci. Chisaka and Capecchi (1991) have described an extensive syndrome of effects arising from homozygosity of a null mutation at *hox-1.5* that results in death shortly after birth. This series of defects is located in a region of the embryo that represents a region of *hox-1.5* expression but does not correlate in any simple way with the expression of *hox-1.5* in general; e.g., normal expression of the gene in lungs, stomach, and spleen is not reflected by a mutant phenotype in these organs. Joyner et al. (1989, 1991) have disrupted the *En-2* gene of mice—a locus that by homology with *Drosophila* was expected to be of great importance in developmental pattern formation—and have demonstrated that the resulting homozygous null mutant mice are phenotypically grossly normal but have a subtle cerebellar alteration. The scope of the phenotypic effect is much less than the field over which the wild-type gene is expressed. These results are unexpected and reemphasize the *experimental* nature of these investigations and their importance in resolution of gene function in mammalian development.

POTENTIAL APPLICATION TO OTHER SPECIES

There has been considerable effort to derive ES cell lines from a number of other species. ES cell lines with the proven ability to produce germ-line transmission from the chimera have so far been described only in mice, although putative ES cell lines have been described in hamster, pig, sheep, and cattle (Doetschman et al. 1988b; Notarianni et al. 1991). To date, none of the putative ES cell lines from other species have demonstrated the capacity to make germ-line chimeras, which is the essential property needed for transgenesis via these cells. In the larger farm animals, there is a considerable time constraint from the life cycle, and in nonmurine experimental animals, the technology of embryo manipulation and transfer still needs to be perfected. We are presently attempting to isolate ES cells from the laboratory rat, the experimental animal of choice for many pharmacological and physiological studies, where much of the experimental embryology is available.

It is expected that mice will remain the premier experimental genetic species, although adapting these techniques to larger domestic farm animals would open the way to a more carefully controlled transgenesis in these species. The results of Petitte et al. (1990), who were able to produce germ-line chimeric chickens by transfer of small numbers of transiently cultured epiblast cells, open the possibility that this technology may become applicable to avian species as well.

TECHNICAL LIMITATIONS AND POSSIBLE RESOLUTIONS

Although ES cell technology is clearly proving to be a very successful method for introducing experimental manipulations of endogenous gene sequences, it is at present laborious and requires expertise in three different laboratory skills: molecular biology, tissue culture, and experimental embryology. Fortunately, refinement of procedures in these areas promises to make the technology much more tractable.

I have previously reviewed the types of constructs that can be used (Evans 1989) and in particular, the technique of positive followed by negative selection for a construct which first goes into a targeted locus and then is removed either by the in-out mechanism described by Valancius and Smithies (1991) and Hasty et al. (1991) or by a second round of targeting where the new homology inserted in the first round is used to remove most of the construct leaving a specific mutation. These techniques lend themselves to construction of genomic clones with a specific series of cassettes all of which are becoming well established.

The second phase—that of ES cell culture and selection or screening of targeted clones—can be very time-consuming, especially when individual clones are grown in a large population prior to screening. Techniques that allow the use of multiwell plates such as those described for freezing (Chan and Evans 1991) and methods under development to allow rapid production of sufficient DNA for Southern blots (coupled with some degree of automated handling) should greatly improve this situation. The technique of immediate PCR analysis of clones (Joyner et al. 1989), or other techniques that can be applied to early screening of clones, will have significant advantages in that there is much less volume of tissue-culture work. In addition, the immediate identification of successfully targeted clones allows a shorter period of cell culture passage, which lessens the chance of loss of the clone's totipotency.

Experimental embryology techniques, including construction of chimeras and test breeding, also present difficulties at the moment. For example, Schwartzberg et al. (1989) reported the relative effectiveness of different host blastocysts. The C57BL/6 inbred mouse line is at the moment the preferred donor of host blastocysts, but unfortunately, these animals have a relatively low fertility rate. Techniques of introducing the ES cells into the embryos at the morula stage rather than blastocyst stage have been investigated with success by some researchers.

It is also clear that the maintenance of totipotentiality of the cells in culture and their ability to produce germ-cell chimeras are of critical importance. Although the various conditions for ES cell growth on STO feeders (Martin and Evans 1975), embryonic fibroblast feeders (Doetschman et al. 1988a), or without feeders but supplemented with either buffalo rat liver (BRL)-conditioned medium or recombinant LIF, have all been satisfactorily used (Smith et al. 1988; Williams et al. 1988), it is clear that any culture crisis may lead to aneuploidy and/or lack of good chimerization potential. Attention to exemplary tissue-culture conditions is clearly vital. Numbers of cell lines have been shown to be effective vectors for germ-line chimerism, and there is little reason to suppose that other such lines are not readily isolated and maintained. It is clear, however, that further improvements to the rigor of the tissue-culture protocols and establishment of well-characterized ES cell stocks will be of great utility. In all, I confidently anticipate that the procedures for the use of ES cells for selective gene modification in mice and other mammals will become much more of a routine process than it is today. It should become the method of choice for experimental genetic study in the complexity of the whole organism.

Acknowledgments

The author is supported by the Wellcome Trust.

References

Bradley, A., M. Evans, M.H. Kaufman, and E. Robertson. 1984. Formation of germ-line chimeras from embryo derived teratocarcinoma cells. *Nature* 309: 225.

Chan, S.Y. and M.J. Evans. 1991. In situ freezing of embryonic stem cells in multiwell plates. *Trends Genet.* 7: 76.

Charron, J., B.A. Malynn, E.J. Robertson, S.P. Goff, and F.W. Alt. 1990. High-frequency disruption of the N-myc gene in embryonic stem and preB cell lines by homologous recombination. *Mol. Cell. Biol.* 10: 1799.

Chisaka, O. and M.R. Capecchi. 1991. Regionally restricted developmental defects resulting from targetted disruption of the mouse homeobox gene hox 1.5. *Nature* 350: 473.

Doetschman, T., N. Maeda, and O. Smithies. 1988a. Targetted mutation of the Hprt gene in mouse embryonic stem cells. *Proc. Natl. Acad. Sci.* 85: 8583.

Doetschman, T., P. Williams, and N. Maeda. 1988b. Establishment of hamster blastocyst-derived embryonic stem (ES) cells. *Dev. Biol.* 127: 224.

Doetschman, T., R.G. Gregg, N. Maeda, M.L. Hooper, D.W. Melton, S. Thompson, and O. Smithies. 1987. Targetted corrections of a mutant HPRT gene in mouse embryonic stem cells. *Nature* 330: 576.

Dyson, P.J., F. Poirier, and R.J. Watson. 1989. Expression of c-*myb* in embryonal carcinoma cells and embryonal stem cells. *Differentiation* **42**: 24.

Evans, M.J. 1989. Potential for genetic manipulation of mammals. *Mol. Biol. Med.* **6**: 557.

Evans, M.J. and M.H. Kaufman. 1981. Establishment in culture of pluripotential cells from mouse embryos. *Nature* **292**: 154.

Hasty, P., R. Ramirez-Solis, R. Krumlauf, and A. Bradley. 1991. Introduction of a subtle mutation into the Hox 2.6 locus in embryonic stem cells. *Nature* **351**: 234.

Hooper, M., K. Hardy, A. Handyside, S. Hunter, and M. Monk. 1987. HPRT-deficient (Lesch-Nyhan) mouse embryos derived from germline colonization by cultured cells. *Nature* **326**: 292.

Joyner, A.L., W.L. Skarnes, and J. Rossant. 1989. Production of a mutation in mouse En-2 gene by homologous recombination in embryonic stem cells. *Nature* **338**: 153.

Joyner, A.L., K. Herrup, B.A. Auerbach, C.A. Davis, and J. Rossant. 1991. Subtle cerebellar phenotype in mice homozygous for a targeted deletion of the En-2 homeobox. *Science* **251**: 1239.

Kim, H.S. and O. Smithies. 1988. Recombinant fragment assay for gene targeting based on polymerase chain reaction. *Nucleic Acids Res.* **16**: 8887.

Kitamura, D., J. Roes, R. Kuhn, and K. Rajewsky. 1991. A B cell-deficient mouse by targeted disruption of the membrane exon of the immunoglobulin µ chain gene. *Nature* **350**: 423.

Koller, B.H., P. Marrach, J.W. Kappler, and O. Smithies. 1990. Normal development of mice deficient in β2M, MHC class I proteins, and CD8+ T cells. *Science* **248**: 1227.

Kuehn, M.R., A. Bradley, E.J. Robertson, and M.J. Evans. 1987. A potential animal model for Lesch-Nyhan syndrome through introduction of HPRT mutations into mice. *Nature* **326**: 295.

Martin, G.R. and M.J. Evans. 1975. Differentiation of clonal lines of teratocarcinoma cells: Formation of embodied bodies in vitro. *Proc. Natl. Acad. Sci.* **72**: 1441.

McKusick, V.A. 1991. Current trends in mapping human genes. *FASEB. J.* **5**: 12.

McMahon, A.P. and A. Bradley. 1990. The Wnt-1 (int-1) proto-oncogene is required for development of a large region of the mouse brain. *Cell* **62**: 1073.

Mucenski, M.L., K. McLain, A.B. Kier, S.H. Swerdlow, C.M. Schreiner, T.A. Miller, D.W. Piefryga, W.J. Scott, and S.S. Potter. 1991. A functional c-*myb* gene is required for normal murine fetal hepatic hematopoeisis. *Cell* **65**: 677.

Notarianni, E., C. Galli, S. Laurie, R.M. Moor, and M.J. Evans. 1991. Derivation of pluripotent, embryonic cell lines from the pig and sheep. *J. Reprod. Fertil.* (suppl.) **43**: 255.

Petitte, J.N., M.E. Clark, G. Liu, G.A. Verrinder, and R.J. Etches. 1990. Production of somatic and germline chimeras in the chicken by transfer of early blastodermal cells. *Development* **108**: 185.

Reid, L.H., E.G. Shesely, H.S. Kim, and O. Smithies. 1991. Contransformation and gene targeting in mouse embyonic stem cells. *Mol. Cell Biol.* **11**: 2769.

Robertson, E.J., A. Bradley, M. Kuehn, and M. Evans. 1986. Germ-line transmission of genes introduced into culture pluripotent cells by retroviral vectors. *Nature* **323**: 445.

Saiki, R.K., S. Scharf, F. Faloona, K.B. Mullis, G. Horn, and H.A. Erlich. 1985. Enzymatic amplification of b-globin sequences and restriction site analysis for diagnosis of sickle cell anemia. *Science* **230:** 1350.

Schwanzberg, P.O., S.P. Goff, and E.J. Robertson. 1989. Germ-line transmission of a C-abl mutation produced by targeted gene disruption of embryonic stem cells. *Science* **246:** 799.

Smith, A.G., J.K. Health, D.D. Donaldson, G.G. Wong, J. Moreau, M. Stahl, and D. Rogers. 1988. Inhibition of pluripotential embryonic stem cell differentiation by purified polypeptides. *Nature* **336:** 688.

Smithies, O., R.G. Gregg, M.A. Boggs, M.A. Koralewski, and R. Kucherlapati. 1985. Insertion of DNA sequences into the human chromosome β-globin locus by homologous recombination. *Nature* **317:** 230.

Soriano, P., C. Montgomery, R. Geske, and A. Bradley. 1991. Targeted disruption of the c-*src* proto-oncogene leads to osteopetrosis in mice. *Cell* **64:** 693.

Stanton, B.R., S.W. Reid, and L.F. Parada. 1990. Germ line transmission of an inactive N-myc allele generated by homologous recombination in mouse embryonic stem cells. *Mol. Cell. Biol.* **10:** 6755.

Thomas, K.R. and M.R. Capecchi. 1990. Targetted disruption of the murine int-1 proto-oncogene resulting in severe abnormalities in midbrain and cerebellar development. *Nature* **346:** 847.

Thomas, K.R., K. Folger, and M.R. Capecchi. 1986. High frequency targeting of genes to specific sites in the mammalian genome. *Cell* **44:** 419.

Valancius, V. and O. Smithies. 1991. Testing an "in-out" targeting procedure for making subtle genomic modifications in mouse embryonic stem cells. *Mol. Cell. Biol.* **11:** 1402.

Williams, R.L., D.J. Hilton, S. Pease, T.A. Willson, C.L. Stewart, D.P. Gearing, E.F. Wagner, D. Metcalf, N.A. Nicola, and N.M. Gough. 1988. Myeloid leukaemia inhibitory factor maintains the developmental potential of embryonic stem cells. *Nature* **336:** 684.

Zilstra, M., M. Bix, N.E. Simister, J.M. Loring, D.H. Raulet, and R. Jaenisch. 1990. β2-microglobulin deficient mice lack CD4$^-$8$^+$ cytolytic T cells. *Nature* **344:** 742.

Zimmer, A. and P. Gruss. 1989. Production of chimaeric mice containing embryonic stem (ES) cells carrying a homobox Hox 1.1. allele mutated by homologous recombination. *Nature* **338:** 150.

Insertional Mutagenesis

Michael A. Rudnicki and Rudolf Jaenisch
Whitehead Institute for Biomedical Research
Cambridge, Massachusetts 02142

The development of techniques that allow the introduction of cloned DNA into the germ line of mice has greatly facilitated the molecular genetic analysis of gene expression and function. The process of generating transgenic mice involves the integration of foreign DNA into host chromosomal sequences, and thus one might expect to disrupt or insertionally mutate host genes at a certain frequency. Unlike spontaneous or radiation-induced mutations, the molecular cloning and analysis of the mutated gene are possible since the presence of the transgene tags the mutated locus. Early experiments revealed the mutagenic potential of this approach, and several fascinating insertional mutants have been described in the last decade. Recently, efforts have been directed toward the development of more systematic strategies for the generation of insertional mutations. In this chapter, we review the methods used to produce insertional mutations, describe examples of such mutants, and discuss the merits of the various approaches.

This chapter discusses:

❏ transgenic techniques used to generate insertional mutants in the mouse

❏ specificity and effect of integration on host genomic sequences

❏ strategies to allow more systematic insertional mutagenesis of the mouse genome

❏ cloning and analysis of chromosomal loci containing an insertional mutation

❏ examples of transgenic mouse strains carrying insertional mutations

INTRODUCTION

Since the early 1980s, the study of mouse genetics has undergone a revolutionary transformation into a powerful tool with which to pursue fundamental questions concerning the biology of mammalian life. This revolution was made possible by the parallel development of molecular biology and of techniques that allow the introduction of foreign genetic information into the germ line of mice. The production of such new mouse strains, termed transgenic mice, has allowed the molecular genetic analysis of many profound biological problems, including embryonic development, oncogenesis, immune function, and regulation of gene expression.

New genetic information was first experimentally introduced into mice by microinjection of purified SV40 DNA into mouse blastocysts (Jaenisch and Mintz 1974). The first germ-line transgenic mouse strains were produced by infection of preimplantation embryos with the Moloney murine leukemia virus (Mo-MLV) (Jaenisch 1976, 1977). However, transgenesis did not become widely applicable until the almost simultaneous technical achievement by several groups whereby germ-line transgenic mice were produced by microinjection of cloned DNA into the male pronucleus of fertilized oocytes (for review, see Jaenisch 1988; Hanahan 1989). More recently, in a similarly important technical advance, embryonic stem (ES) cells have been used as the means to introduce genes into the germ line or to target specific endogenous genes for disruption by homologous recombination (Capecchi 1989: Rossant and Joyner 1989; see Evans, this volume).

It soon became apparent that the introduction of foreign DNA into the mouse would occasionally generate recessive mutations. These mutations were often a consequence of integration of the foreign gene into or near some host gene. Molecular genetic analysis of this type of mutation is possible since the transgene (foreign DNA) can be used as a probe to clone the interrupted gene, which then allows the structure and function of the mutated gene to be analyzed with all the tools of modern molecular biology. In contrast, identification and cloning of spontaneous or radiation-induced mouse mutations are extremely difficult. The attraction of insertional mutagenesis is the possibility of discovering novel genes and hence novel paradigms that may regulate complex biological phenomena. This approach has proven to be essential in the molecular genetic study of such diverse organisms as yeast (Roeder and Fink 1980), maize (Federoff 1983), and *Drosophila* (Spadling and Rubin 1983).

Since the advent of mouse transgenesis, there have been number of reports describing new mutations due to the insertion of either retroviral or foreign DNA into a chromosomal locus. Several of these interrupted loci have been cloned, but only a few have been subject to extensive analysis. The introduction of the ES cell system has allowed the development of strategies that show much promise for more systematic generation of insertional mutants. In this chapter, we discuss established and new strategies for the generation of insertional mutations and describe examples generated by these procedures.

STATEGIES OF INSERTIONAL MUTAGENESIS

Insertional mutants have been produced by a variety of techniques, including exposure of either pre- or postimplantation embryos to infectious retrovirus, microinjection of cloned DNA into the male pronucleus of fertilized oocytes, and, more recently, retroviral infection or DNA transformation of cultured ES cells (summarized in Fig. 1). In this section, we describe the various approaches used to generate insertional mutants.

Retroviral infection of embryos

Retroviruses are RNA viruses that replicate via a DNA intermediate called the provirus (for review, see Varmus 1988). Transgenic mouse strains can be generated at high efficiency by retroviral infection of preimplantation embryos (for review, see Jaenisch 1988). These experiments most often make use of replication-competent or recombinant derivatives of the Mo-MLV (for review, see Weiss et al. 1984). Briefly, 8–16-cell embryos are exposed to concentrated retrovirus or are cocultured with a layer of virus-producing cells prior to reintroducing the infected embryos into the uteri of pseudopregnant mice. Alternatively, concentrated retrovirus is microinjected into the blastocoelic cavity of blastocyst-stage embryos (Fig. 1). Approximately 80% of the preimplantation embryos exposed to infectious Mo-MLV in this fashion give rise to transgenic animals and almost all of these are capable of transmitting the integrated provirus through their germ line. Most of these mice are mosaic for one or more proviruses that usually genetically segregate in the first generation (Jaenisch 1976; Jaenisch et al. 1981; Soriano and Jaenisch 1986). Concentrated retrovirus can also be injected into the amniotic cavity of the 8-day-old postimplantation embryo (Jaenisch 1980; Jaenisch et al. 1981; Stuhlmann et al. 1984). However, the frequency of germ-line transmission of integrated provirus of mice infected at these later stages is greatly reduced. About 5% of the transgenic mice generated by retroviral infection of early mouse embryos exhibit mutant phenotypes when bred to homozygosity (for review, see Gridley et al. 1987). All of the insertional mutations generated by proviral integration have been recessive and most of these have been lethal (see Table 2).

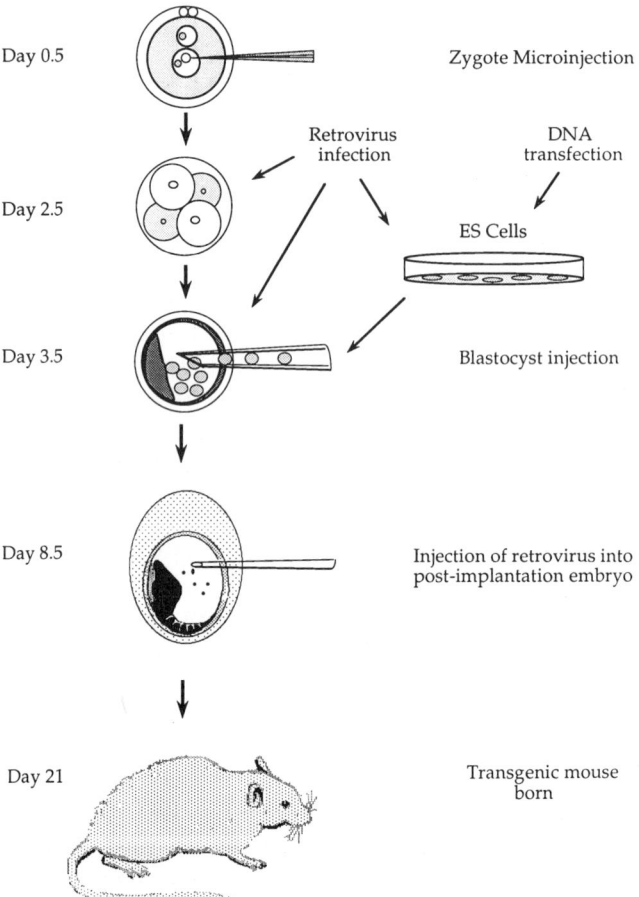

Figure 1 Introduction of new genetic information into the germ line of mice. (*Day 0.5*) Cloned DNA is microinjected into the male pronucleus of fertilized oocytes. (*Day 2.5*) Cultured preimplantation embryos are exposed to retrovirus. (*Day 3.5*) Genetically modified ES cells are injected into blastocyst-stage embryos, or concentrated retrovirus is injected into the blastocoel. (*Day 8.5*) Concentrated retrovirus is injected into the amniotic cavity of postimplantation embryos. (*Day 21*) Transgenic mouse is born and subsequently bred to test for germ-line transmission of the transgene.

In early experiments, replication-competent Mo-MLV was used to infect embryos (Jaenisch 1976; Jaenisch et al. 1981). However, replication of this type of retrovirus leads to viremia, which often induces leukemia in the mouse. More recently, defective or replication-incompetent recombinant retroviruses can be produced in high titers from packaging cell lines in which the formation of wild-type virus is very rare (Mann et al. 1983).

Insertional mutations can also be generated by endogenous retroviruses, and two "classical" mouse mutations have been shown to be caused by proviral insertion into a host gene (Jenkins and Copeland 1985; Stoye et al. 1988). Replication of endogenous retroviruses can be induced experimentally by interbreeding certain mouse strains. When *SWR/J-RF/J* mice were bred to ecotropic virus-negative *SWR/J* mice, about 19% of the offspring acquired new proviral genomes. In one experiment, 1 of 18 newly acquired proviruses was found to cause a recessive lethal insertional mutation (Spence et al. 1989). Thus, this approach may be a simpler alternative to the experimental production of insertional mutants by retroviral infection of preimplantation embryos.

Microinjection of DNA into zygotes

Microinjection of cloned DNA into the pronucleus of fertilized mouse oocytes results in chromosomal integration of the injected DNA in about 25% of the animals born. Approximately 20% of these founder animals are mosaic for the presence of the transgene, suggesting that, in these cases, integration occurred after the first cleavage (Gordon et al. 1980; Brinster et al. 1981; Costantini and Lacy 1981; Gordon and Ruddle 1981; E. Wagner et al. 1981; T.E. Wagner et al. 1981). Pronuclear injection of recombinant DNA has proven to be an efficient procedure for the production of new mouse strains that express the introduced genetic information in a predictable and tissue-specific manner (for review, see Jaenisch 1988; Hanahan 1989; for a description of the zygote microinjection technique, see Hogan et al. 1986). Insertional mutants are generated at a frequency of about 8%, and many of these are recessive lethal mutants (for review, see Palmiter and Brinster 1986). However, this frequency may be a low estimate, because many phenotypes may be too subtle to be detected easily.

The majority of transgenic mice generated by zygote microinjection contain multiple copies of the introduced DNA, most often arranged as tandem repeats at a single locus (Brinster et al. 1981; Costantini and Lacy 1981). It is believed that the linearized injected DNA first associates via the ends to form long head-to-tail arrays, which then integrate in a nonhomologous fashion into chromosomal DNA. There is no evidence for specificity or preference for the integration of transgenes (Lacy et al. 1983; Palmiter and Brinster 1986), but some sites may be more accessible than others (e.g., the *ld* locus, see below). The host chromosomal sequences at the integration site often contain deletions, inversions, duplications, translocations, and other modifications (Table 1).

Most of the insertional mutants that have been discovered in transgenic strains were generated by pronuclear microinjection of cloned DNA into zygotes (see Table 2). This is due to the large number of transgenic strains that have been produced by this procedure, rather than by any inherent advantage of this method for generating insertional mutants.

Table 1 Integration of retroviruses and microinjected DNA

Method	Copy number	Effect on integration site	Integration site bias
Retroviral infection	single provirus	short duplication on either side of provirus	integration occurs near DNase-I-hypersensitive regions perhaps preference for transcriptionally active chromatin
Pronuclear microinjection	often tandem arrays of many copies	often have complex rearrangements: deletions, inversions, duplications, translocations of both foreign and host sequences	unknown; perhaps some sites more accessible than others

Transfection and infection of ES cells

ES cells are continuous cell lines isolated from mouse blastocyst-stage embryos (Evans and Kaufman 1981; Martin 1981). These cells are completely equivalent to the pluripotent cells of the early embryo and can contribute to all cell lineages, including the germ line, when microinjected into blastocyst-stage embryos (Bradley et al. 1984). Thus, in this procedure, genetically modified ES cells can be microinjected into blastocysts, the injected embryos transferred into the uteri of pseudopregnant mice, and the resulting chimeric mice bred to produce new mouse strains containing the genetic modification of interest (Fig. 1) (for descriptions of these techniques, see Hogan et al. 1986; Robertson 1987).

ES cells can be genetically modified in tissue culture by a variety of approaches. Cloned DNA can be transfected (Gossler et al. 1986), lipofected (M.A. Rudnicki and R. Jaenisch, unpubl.), or electroporated (Thomas and Capecchi 1987) into ES cells, and stable transformants can be derived using standard antibiotic selection regimes (see Robertson 1987). Alternatively, ES cells can be infected with recombinant retrovirus (summarized in Fig. 1) (Kuehn et al. 1987).

A very exciting application of this new technology is the targeted disruption of endogenous genes by homologous recombination in ES cells (for review, see Capecchi 1989; Rossant and Joyner 1989; see Evans, this volume). This approach can result in the disruption or modification of any cloned gene in the mouse. In addition, the ES cell system also holds much promise for the systematic production of insertional mutants in the mouse.

COMPARISON OF APPROACHES TO PRODUCE INSERTIONAL MUTATIONS

Many of the best-studied insertional mutations were generated accidentally in the course of producing transgenic mice for some other purpose. The type of approach used can have different consequences on the insertion site of the transgene, the clonability of the mutated locus, and the frequency of insertional mutagenesis. In this section, we discuss the advantages and disadvantages of using the different procedures to generate insertional mutants (see Table 1).

Retrovirus insertional mutagenesis

Germ-line integration of retroviruses has yielded several insertional mutations in the mouse (see below). Retroviral replication results in the integration of a single provirus in host chromosomal DNA. In contrast to the insertion site of microinjected DNA, the only rearrangement at a Mo-MLV proviral integration site is a short duplication on either side of the provirus (Table 1). Therefore, the mutagenic lesion is likely to be a direct consequence of proviral integration into or near an endogenous gene. The presence of a single provirus and the absence of rearrangements facilitate the molecular cloning and analysis of chromosomal sequences flanking the proviral integration sites.

Retroviruses appear to integrate preferentially near DNase-I-hypersensitive sites in host chromatin, suggesting a preference for integration into transcriptionally active chromatin (Vijaya et al. 1986; Rohdewohld et al. 1987; Shih et al. 1988). Thus, retroviral integration might occur preferentially in genes that are transcribed in embryonic cells. If this is correct, then only a subset of all genes may be potentially mutable by retroviral integration.

Molecular characterization of the mutation can be further facilitated by inclusion of bacterial suppressor tRNA genes to allow selective cloning of the provirus and surrounding host chromosomal sequences in bacteriophage λ (Reik et al. 1985; Soriano et al. 1987). Development of shuttle vectors that are designed to allow rapid recovery and cloning of chromosomal sequences flanking the integration site may also prove useful (Cepko et al. 1984; Berger and Bernstein 1985). However, with the widespread adoption of the polymerase chain reaction (PCR; Erlich 1989), it is now very simple to amplify sequences adjacent to the proviral integration site directly, and then use the amplified DNA as molecular probes (Silver and Keerikatte 1989).

Insertional mutagenesis by microinjection of zygotes

Several fascinating insertional mutants have been generated by zygote microinjection (Table 2; see below). Microinjected transgenes can serve as molecular entry points into different regions of the mouse genome and provide useful molecular genetic markers. However, production and analysis of insertional mutations by this approach have a number of serious limitations. The major problem is the integrity of the insertion site. The molecular cloning and analysis of the mutated gene are only possible if the foreign DNA has integrated into or near the mutated gene. In many cases, however, large inversions or translocations may result in the disruption of more than one gene, making it difficult to identify the mutation responsible for the resulting phenotype. In addition, the large copy numbers of the introduced DNA and the complex rearrangements in and about the integration site can make molecular analysis of insertional mutants a frightful enterprise. To date, the cDNA cloning of transcripts from genes disrupted by the presence of microinjected foreign DNA has been reported in only a single case, reflecting the difficulty of the molecular analysis of mutations generated by this method. Furthermore, characterization of the mutant phenotype may be complicated by the expression of a dominant-acting transgene.

Insertional mutagenesis in ES cells

The use of the ES cell system to generate insertional mutations has several practical and theoretical advantages over the approaches discussed previously. The primary advantage of the ES system is that ES cells are grown in tissue culture and thus large numbers of clones can be analyzed for gene disruption events. In addition, recombinant vectors can be designed such that the recovery of ES cell clones containing insertions into genes is enriched or selected for in vitro before attempting to introduce the mutations into mice. These vectors typically contain a β-galactosidase (*lacZ*) gene and are designed such that *lacZ* expression reflects the expression pattern of the interrupted gene. Histological determination of the developmental and spatial pattern of tissue-specific expression can be directly discerned (see, e.g., Kotheray et al. 1988). The possibility of prescreening ES cell lines that contain these vectors in vitro and in the founder chimera before germ-line transmission allows for a more efficient and rational analysis of mutations than is possible by alternative approaches (see below). In addition, dominant lethal mutations can be recovered since the integration site is not lost with the death of the founder animal.

Gossler et al. (1989) described the first vectors designed specifically to disrupt endogenous genes in ES cells (for review, see Rossant and Joyner 1989). These vectors are designated enhancer-trap and gene-trap, and contain *lacZ* and neomycin phosphotransferase (*neo*) encoding sequences (Fig. 2a,b). The enhancer-trap vector contains a minimal promoter, too weak to stimulate transcription alone, upstream of the *lacZ*-coding sequences. The gene-trap vector contains a splice acceptor sequence just upstream of the *lacZ*-coding region and lacks a translation start site. In both vectors, selection of stable transformants is facilitated by the presence of a *neo* gene under control of a constitutive promoter, downstream from the *lacZ* sequences. After electroporation of these vectors into ES cells, large numbers of G418-resistant transformants can be screened for *lacZ* expression. In transformants containing enhancer-trap vectors, *lacZ* expression will be detectable if integration is near *cis*-acting regulatory elements that activate the minimal promoter. In transformants containing gene-trap vectors, *lacZ* expression will be detected only after integration in the correct orientation into an intron of a protein-encoding gene, where fusion occurs in-frame with *lacZ*. These vectors also contain the bacterial supIII$^+$ suppressor tRNA gene to aid the molecular cloning of the interrupted loci.

In undifferentiated ES cell clones electroporated with these vectors, about 10% of the enhancer-trap vector transformants and 2% of the gene-trap vector transformants were found to express *lacZ*. Chimeric mice produced by microinjection of these ES cell lines into blastocysts exhibited widely different tissue-specific patterns of *lacZ* expression that were characteristic of an individual ES clone. It has not been reported what proportion of these integration events have disrupted an endogenous gene. This analysis requires germ-line transmission and breeding of mice that are homozygous at the integration site. An issue yet to be resolved is the integrity of the insertion site in electroporated ES cell lines. If inversions, deletions, translocations, and other rearrangements are commonly found about the insertion vector, then the molecular cloning of the insertion site may become difficult.

Recombinant retroviruses can infect ES cells (Robertson et al. 1986) and can be used to generate insertional mutations in an approach analogous to the gene-trap stategy. The Mo-MLV proviral promoter is found within the 5' long terminal repeat (LTR) and is not capable of stimulating transcription within undifferentiated stem cells (Weiher et al. 1987). Thus, in embryonal carcinoma (EC) cells (similar to ES cells; see Rudnicki and McBurney 1987), a recombinant retrovirus containing *neo*-coding sequences will function as a promoter-trap because *neo* expression will only occur when the provirus has integrated near a strong cellular promoter or splice donor (see MP10 in Fig. 2c) (Barklis et al. 1986; Peckham et al. 1989). Transgenic mice have been produced in our laboratory from ES cells infected with this virus, and two of six cell lines gave rise to novel recessive insertional mutants (E. Li and R. Jaenisch, unpubl.). Similarly, 3 of 13 transgenic mouse strains produced from ES cells infected with another recombinant retrovirus (see below) display recessive mutant phenotypes (P. Soriano, pers. comm.). These preliminary results suggest that this approach may give rise to insertional mutants at high frequencies.

a. Enhancer-trap

b. Gene-trap

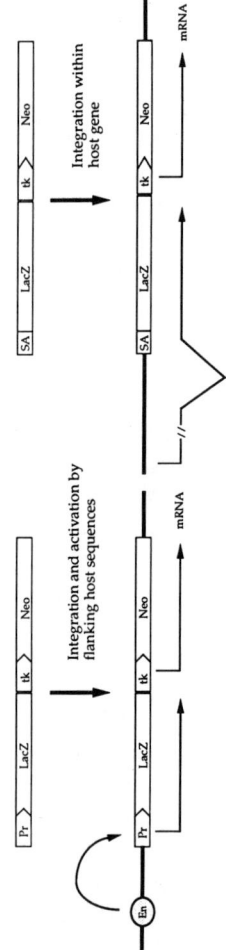

c. Promoter-trap; Mp10 provirus

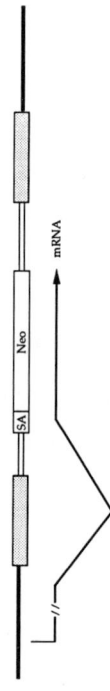

d. Promoter-trap; U3His provirus

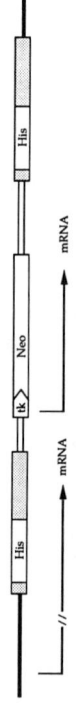

Figure 2 Design of vectors that enrich for insertional mutagenesis after introduction into ES cells. (*a*) Activation of an enhancer-trap vector occurs after integration near an activating element (En) in flanking host chromosomal sequences. *lacZ* is transcribed from the activated minimal promoter (Pr). (*b*) Activation of a gene-trap vector occurs after integration within an expressed host gene. The *lacZ* gene lacks a translational start site and is expressed only when fused in-frame with a host transcription unit (Gossler et al. 1989; for review, see Rossant and Joyner 1989). The retroviruses Mp10 (*c*) (Barklis et al. 1986) and U3His (*d*) (von Melchner and Ruley 1989) function as promoter traps in ES cells because antibiotic resistance will occur only when transcription originates from host chromosomal sequences. (SA) Splice acceptor sequence; (tk-Neo) thymidine kinase constitutive promoter expressing neomycin phosphotransferase; (His) histidinol dehydrogenase. (*Solid bar*) Host chromosomal sequences; (*stippled boxes*) retroviral LTRs.

In alternative designs, recombinant Mo-MLV retroviruses have been built to function specifically as promoter traps and could be used to generate insertional mutations in ES cells with high frequency. In one design, the coding sequences of the selectable marker histidinol dehydrogenase (*his*) were inserted into the LTR such that proviral integration placed the *his* gene 30 bp from the flanking cellular sequences (Fig. 2d) (von Melchner and Ruley 1989). In this fashion, *his* resistance would only occur when transcription originated from promoters located in flanking chromosomal DNA, and indeed, several strong promoters have been cloned using PCR-based amplification (von Melchner et al. 1990). In another retroviral vector design, *lacZ*-coding sequences were placed in the antisense orientation relative to the Mo-MLV LTRs. Expression of the *lacZ* protein only occurs when the provirus integrates downstream from a cellular promoter in an orientation opposite to that of the gene (P. Soriano, pers. comm.). In the future, these kinds of retroviral vectors, combined with the use of the ES cell system, will allow more systematic generation of insertional mutants in the mouse.

EXAMPLES OF INSERTIONAL MUTATIONS

The phenotypic and molecular analysis of insertional mutants often represents a daunting task that can require a long-term effort. Although numerous insertional mutations have been described over the last 10 years, in only a few cases has our understanding of the molecular basis of the phenotypes been advanced. This section describes a number of mutant mouse strains that have been derived by the different approaches detailed above. Some of the mutations that have had extensive molecular analyses are described, followed by a discussion of several mutations for for which only a limited understanding of the mutant gene is available. A list of these findings and some additional mutations are summarized in Table 2.

Molecularly characterized insertional mutations

Mov13, *disruption of the* Col1a1 *gene* The first insertional mutant described (Jaenisch et al. 1983) was the *Mov13* mouse strain, derived after exposing postimplantation mouse embryos (day 8) to Mo-MLV. Embryos homozygous at the *Mov13* locus develop necrosis of erythropoietic and mesenchymal cells in the liver, which is followed by vascular rupture and sudden death at day 13 of gestation, presumably caused by failure of blood circulation (Lohler et al. 1984).

The *Mov13* locus was cloned using the Mo-MLV provirus as a probe, and molecular analysis indicated that the provirus had integrated in the first intron of the α1(I) collagen (*Col1a1*) gene in the opposite

Table 2 Mouse insertional mutations

Transgenic strain	Chromosomal location	Cloning status	Procedure	Homozygous phenotype	References
Characterized mutants					
Mov13	Col1a1	G, cDNA	RV, E	midgestation embryonic lethal	Jaenisch et al. (1983)
Mov34	Chr 8	G, cDNA	RV, E	postimplantation embryonic lethal	Soriano et al. (1987)
Mpv17	Chr 5	G, cDNA	RV, E	adult lethal nephrotic syndrome	Weiher et al. (1990)
Mpv20	Chr 9	G, cDNA	RV, E	preimplantation embryonic lethal	D. Gray and R. Jaenisch (unpubl.)
Dilute	dilute (*d*)	G, cDNA	EndRV	washed-out coat color	Jenkins et al. (1981)
161a-1	limb def. (*ld*)	G, cDNA	M	limb deformities, renal aplasias	Woychik et al. (1985)
HPRT⁻	hprt	d	RV, ES	reduced brain dopamine, Lesch-Nyhan syndrome	Kuehn et al. (1987)
Tg6208	steel (*sl*)	d	M	lethal defects in stem cell lineages	Keller et al. (1990)
Incompletely characterized mutants					
Hβ58	Chr 10	G, Pr	M	postimplantation lethal	Radice et al. (1991)
pHT1-1	legless (*lgl*)	G	M	skeletal, craniofacial, visceral deformities	McNeish et al. (1988)

INSERTIONAL MUTAGENESIS 25

Adp	Acrodysplasia	G	M	paternally dominant limb deformity	D. Solter (pers. comm.)
358-3	Chr 13 (add)	G	M	deformed phalanges of forelimb	Pohl et al. (1990)
OVE3A	Chr 14 (sys)	G	M	male sterility	MacGregor et al. (1990)
OVE1B	downless (dl)	G	M	absence of guard hairs	Shawlot et al. (1989)
Line A	pygmy (pg)	G	M	dwarfism	Xiang et al. (1990)
p447	hotfoot (ho)	G	M	motor disorder, male sterility	Gordon et al. (1990)
p432	Purk. cell. deg.	G	M	neuronal degeneration, defective sperm	Krulewski et al. (1989)
Tg4	dystonia (dt)	n.c.	M	lethal motor neuron degeneration	Kotheray et al. (1988)
413	unknown	n.c.	M	postimplantation lethal	M. Kuehn (pers. comm.)
Hairless	hairless (hr)	G	RV, ES	hair loss	Stoye et al. (1988)
Srev-5	Srev-5	G	EndRV	embryonic lethal	Spence et al. (1989)
RSV-CAT/2	unknown	n.c.	EndRV	fusion of toes in paws	Overbeek et al. (1986)
HUGH/3	unknown	n.c.	M	postimplantation embryonic lethal	Wagner et al. (1983)
HUGH/4	unknown	n.c.	M	postimplantation embryonic lethal	Wagner et al. (1983)
CV4	unknown	n.c.	M	postimplantation embryonic lethal	Shani (1986)
Tg.ple	Chr 15 (ple)	n.c.	M	perinatal lethal	Beier et al. (1989)

Abbreviations: (Chr) Chromosome; (G) genomic locus cloned; (cDNA) cDNA cloned; (d) disruption of previously characterized genes; (n.c.) not cloned; (Pr) transcripts detected with conserved flanking probe; (RV) retrovirus infection; (E) embyro; (EndRV) endogenous retrovirus; (M) microinjection; (ES) embryonic stem cells.

orientation (Schnieke et al. 1983; Harbers et al. 1984). The presence of the provirus at the *Mov13* locus prevents the appearance of transcription-associated DNase-I-hypersensitive sites in chromatin upstream of the *Col1a1* promoter (Breindl et al. 1984) and results in de novo methylation of collagen sequences surrounding the integration site (Jahner and Jaenisch 1985). Transcription of the *Col1a1* gene in *Mov13* mice is reduced 20-200-fold in most (Hartung et al. 1986), but not all, tissues (Kratochwil et al. 1989). This suggests that proviruses can affect gene expression in a tissue-specific manner as has been seen for retrotransposons in *Drosophila* (Bingham and Zacher 1989; Boeke 1989) and in the activation of proto-oncogenes by murine retroviruses (Varmus and Brown 1989). In vitro, no detectable *Col1a1* transcription occurs after transfection of the cloned *Mov13* locus containing either the intact retrovirus or a single LTR into fibroblast cells (D.D. Barker et al., in prep.). These results suggest that the mutation is not due to altered methylation or chromatin structure, but rather to the displacement or disruption of *cis*-acting regulatory sequences within the first intron.

Mov34, *early embryonic lethal phenotype* Infection of preimplantation embryos with a recombinant replication-competent Mo-MLV, containing a bacterial *supF* gene within each LTR (Reik et al. 1985; Soriano and Jaenisch 1986), was used to generate a number of transgenic mouse strains. One of these, *Mov34*, exhibits a recessive lethal phenotype. Embryos homozygous at the *Mov34* locus develop normally into blastocysts but die soon after implantation, before reaching the egg cylinder stage. The chromosomal locus containing the proviral integration site was cloned and analyzed. The provirus maps in the same orientation and to the 5' side of a gene that expresses an abundant and ubiquitous 1.7-kb transcript (Soriano et al. 1987) encoding a protein conserved within mice, humans, and *Drosophila*. The *Mov34* locus is located on mouse chromosome 8, the human homolog has been mapped to chromosomal region 16q23-q24, and the *Drosophila* homolog has been mapped to 60B,C on chromosome 2 (Gridley et al. 1990). The murine and *Drosophila* proteins share 62% identity at the amino acid level and have no significant homology with proteins of known function. The murine nucleotide sequence indicates that the protein has a molecular weight of 39,000 and a very hydrophilic carboxyl terminus consisting of alternating positive and negative charges (Gridley et al. 1990).

Mpv17, *adult lethal nephrotic syndrome* The *Mpv17* mouse strain was derived after preimplantation embryos were exposed to the recombinant retrovirus MPSV*neo* (Seliger et al. 1986; Weiher et al. 1987). Mice homozygous at the *Mpv17* locus display an adult lethal phenotype and die of kidney failure between 2 and 9 months of age (Weiher et al. 1990). The disease first manifests itself in an increase of blood cholesterol levels, followed by elevation in blood urea nitrogen and creatinine and a decrease in serum albumin and hemoglobin. Death of the animal occurs shortly thereafter. Hematological analysis indicates severe normocytic and normochromic anemia. These symptoms, particularly the proteinuria, hypoalbuminemia, and hyperlipidemia, are typical of nephrotic syndrome in humans. Microscopic examination of the kidneys reveals damage at the level of the glomerulus, specifically focal and segmental glomerulosclerosis with accumulation of hyaline material.

The chromosomal locus containing the *Mpv17* integration site was cloned, and the *Mpv17* provirus was found to have disrupted a conserved gene that encodes a ubiquitous 1.7-kb mRNA. In mice homozygous for the *Mpv17* locus, no such transcript is detected. The nucleotide sequence of the mRNA indicates that the protein has 176 amino acids and contains two hydrophobic domains, which may suggest an association with a membrane system. It has no significant homology with proteins of known function.

Dilute* (d) *locus The dilute (*d*) coat color locus on chromosome 9 has been defined by more than 200 spontaneous and induced recessive mutations (for review, see Rinchik et al. 1985). Mice homozygous at the *d* locus exhibit a "washed-out" coat color as a consequence of an abnormal adendritic morphology of neural-crest-derived melanocytes. Some *d* mutations exhibit additional phenotypes including lethal neurological opisthotonus characterized by seizures and followed by death at about 3 weeks of age. These observations suggested roles for the *d* gene in melanocyte and neuronal cell function.

Molecular cloning of the *d* locus was possible after it was found that an endogenous ecotropic murine leukemia virus (Emv-3) had insertionally mutated the *d* locus in one of the old *d* mutations of the mouse fancy (Jenkins et al. 1981). Molecular analysis of the regions flanking the proviral integration site generated probes that have allowed the elucidation of the genomic organization of spontaneous and radiation-induced *d* alleles and the identification of *d* transcripts (Strobel et al. 1990). The *d* cDNA has been cloned and the sequence indicates that the transcript encodes a 1852-amino-acid protein that appears to be a novel type of myosin heavy chain (Mercer et al. 1991).

Limb deformity* (ld) *locus The transgenic mouse strain *165-1a* (S line), generated after pronuclear microinjection of a mouse mammary tumor virus (MMTV)-*myc* fusion construct, exhibits defective limb development in homozygous animals. The insertion site is located on chromosome 2 and appears to be allelic with two other recessive *ld* mutations. The mutant phenotype is associated with synostosis of the long bones of all limbs, oligodactyly and syndactyly of the bones of the paw plates, and renal aplasia. Molecular cloning and analysis of the locus revealed five copies of the transgene and a 1-kb deletion of host chromosomal sequences at the integration site (Stewart et al. 1984; Woychik et al. 1985).

Analysis of the sequences flanking the integration site has identified an evolutionarily conserved gene from which are transcribed a complex set of mRNAs encoding a set of related proteins named formins. These proteins are translated from multiple transcripts generated by alternative splicing and use of different polyadenylation sites (Woychik et al. 1990). Three major species of transcripts (13, 7, and 5 kb), as well as various smaller transcripts, are expressed widely in a complex pattern in many tissues throughout development. The three major formin transcripts are absent in mice containing two independent *ld* alleles, confirming that this mutation results in a loss of function (Maas et al. 1990). The large mRNA species encodes a 1468-amino-acid protein with a predicted molecular weight of 164,000. Various other protein products can be predicted depending on the splicing pattern. The amino acid sequence has no significant homology with proteins of known function, and lacks any obvious signal peptide, transmembrane domain, or DNA-binding domain. The only notable features are a basic amino terminus, an α-helical domain near the carboxyl terminus, and a proline-rich core.

Examination of the early morphological development of the limb buds in transgenic *ld* embryos has demonstrated that the apical ectodermal ridge fails to differentiate properly at day 10 of gestation (Zeller et al. 1990). This structure is thought to be important in providing positional information to the developing limb bud. In the wild-type limb bud, the apical ectoderm was found to express levels of formin mRNA about fivefold higher than the underlying mesenchymal tissues. These results suggest a role for formins in limb development. The functions of the various isoforms of formins in different tissues during embryonic development remain to be elucidated. Recently, an independently derived transgenic mouse strain was found to contain an insertionally mutated *ld* locus. This may suggest that the *ld* locus has a high susceptiblity to insertional mutagenesis after pronuclear injection of DNA (A. Messing et al., pers. comm.).

Mouse model of Lesch-Nyhan syndrome In humans, deficiency of the purine salvage enzyme, hypoxanthine phosphoribosyl transferase (HPRT), causes severe neurological and behavioral defects. To create a mouse model of this disease, ES cells were infected with defective Mo-MLV retrovirus and selected in HAT (hypoxanthine, aminopterin, thymidine) medium for cells lacking HPRT activity. In humans and mouse, the *hprt* gene is located on the X chromosome. Thus, in male ES cells, inactivation of the *hprt* gene requires only a single proviral integration into the X-chromosome-linked locus. ES cells lacking HPRT activity were cloned (the presence of a provirus within the *hprt* gene was verified) and injected into blastocysts to generate mutant mice. Hemizygous males or homozygous females lack detectable HPRT activity, yet appear to be normal (Keuhn et al. 1987). In contrast to the human disease, these animals exhibit no self-mutilation and have no motor impairments in tests of basal ganglia function. However, mutant mice do have reduced levels of dopamine in the striatum (Finger et al. 1988). Analysis of monoamine deficiency in these mice revealed less drastic changes than in Lesch-Nyhan patients (Dunnett et al. 1989). These results suggest that mice lacking HPRT activity are able to compensate, perhaps by utilizing alternative biochemical pathways to maintain physiological levels of purines and neurotransmitters. Although HPRT null mice do not phenotypically resemble the Lesch-Nyhan syndrome, they are useful in understanding differences in nucleic acid metabolism between mice and humans.

Incompletely characterized insertional mutations

Hβ58, *early postimplantation lethal phenotype* The *Hβ58* transgenic mouse strain was produced by pronuclear microinjection of the cloned human β-globin gene (Radice et al. 1991). Homozygous *Hβ58* mice develop abnormally after the primitive streak stage (day 7.5) and die at about day 10. The developmental anomalies include a retardation in the growth of the embryonic ectoderm and abnormalities in the amnion and chorion. The transgene insertion site was molecularly cloned and analyzed. The integration site is located on chromosome 10, contains 10–20 copies of a human β-globin transgene, a deletion of 2–3 kb of host DNA, and an insertion of mouse repetitive DNA of unknown origin. Evolutionarily conserved regions flanking the insertion site have been identified, and one of these hybridizes to a 2.7-kb mRNA expressed in 8.5-day embryos.

Legless* (lgl) *mutation The transgenic mouse strain *pHT1-1* was generated after pronuclear DNA microinjection. It contains about 25 copies of tandemly repeated transgene constructs. Mice homozygous at the *pHT1-1* insertion locus display a recessive perinatal lethal phenotype associated with a variable range of unique limb and craniofacial defects (McNeish et al. 1988). Homozygous mutant mice lack distal hindlimb structures and often lack forelimb structures. They also show craniofacial deformities ranging from mild to severe clefts of the face and palate, abnormalities in brain development, and transposition of internal organs (McNeish et al. 1990). This mutant phenotype does not resemble any known mutant mouse strains. The molecular cloning of the transgene locus has resulted in the identification of evolutionarily conserved flanking sequences and proximity to a CG island (S. Potter, pers. comm.).

Dominant Acrodysplasia* (Adp), *an imprinted locus The *Adp* transgenic strain of mice was generated by zygote microinjection of a bovine papillomavirus (BPV)–metallothionein (MT)–growth hormone fusion construct (D. Solter, pers. comm.). Mice that have inherited the transgene paternally display a phenotype characterized by loss of distal limb structures. However, the mutation shows incomplete penetrance and defects ranging from complete absence of digits to mild deformities of digits. Interestingly, the BPV transgene is expressed only in affected mice, and these mice go on to develop papillomas at about 6 months of age. In contrast, mice who have maternally inherited the transgene are completely normal. These results suggest a modified dominant pattern of inheritance, possibly the consequence of imprinting at the genomic locus containing the transgene. Imprinting of transgenes, where expression depends on inheritance, has been described in several instances (for review, see Surani et al. 1988).

Anterior digit-pattern deformity* (add) *mutation The transgenic mouse strain *358-3* was generated by pronuclear microinjection (Dente et al. 1988). Homozygous *358-3* mice display a forelimb defect where the thumb becomes elongated and the second digit often contains an extra phalanx (Pohl et al. 1990). The chromosomal locus containing the transgene insertion site has been molecularly cloned and analyzed. The integration site contains several copies of the transgene as well as some unknown mouse DNA. The transgene is located close to the centromere of chromosome 13 and maps near the genetic locus of the known mutant extra-toes (*Xt*). Genetic data and molecular analysis with flanking probes support the hypothesis that *add* and *Xt* mutations are allelic.

Symplastic spermatids* (sys), *defective spermatogenesis The transgenic mouse strain *OVE3A* was generated by pronuclear microinjection and contains about five to ten copies of a Rous sarcoma virus (RSV) *lacZ* reporter transgene (MacGregor et al. 1990). Homozygous male *OVE3A* mice are sterile, whereas females are completely fertile. In the testes of homozygous males, the developing spermatids do not complete maturation but form prominent multinucleated syncytia. In addition, the sertoli cells often contain abnormal cytoplasmic vacuoles. The chromosomal locus containing the transgene insertion site has been molecularly cloned and has been located to chromosome 14, about 4 cM proximal to the esterase-10 gene.

Downless* (dl) *locus The transgenic mouse strain *OVE1B* was generated by zygote microinjection of an α-crystallin/SV40 fusion construct and contains a single truncated copy of the construct (Shawlot et al. 1989). Homozygous *OVE1B* mice have thin greasy hair with bald patches behind the ear, lack certain glands, guard hairs, and hair on the tail and display a kink at the tip of the tail. The chromosomal locus containing the transgene integration site was molecularly cloned and has been located by molecular and genetic means to chromosome 10. Analysis of the integration site suggests that a large deletion of endogenous sequences occurred during integration of the transgene. Breeding and mapping evidence suggests that the mutation in transgenic mouse strain *OVE1B* is allelic with the known mouse mutation downless.

Pygmy* (pg) *locus The trangenic mouse strain *Line A* was produced by pronuclear injection and contains a human β-globin transgene (Xiang et al. 1990). Mice homozygous at the transgene locus are reduced in size to about 40% of normal weight, yet have normal levels of growth hormone and somatostatin. The integration site has been cloned and analyzed. Molecular and genetic evidence suggests that the mutation in *Line A* is allelic with the known mouse mutant pygmy.

Hotfoot* (ho) *locus The transgenic mouse strain *p447* was generated by zygote microinjection and contains several copies of an SV40-dihydrofolate reductase (DHFR) transgene (Gordon et al. 1990). Mice homozygous at the transgene locus display motor disorder characterized initially at 14 days of age by an ataxic gait, which progresses by 25 days into jerky flexions of the limbs in an alternating stereotypic fashion. An additional symptom is male sterility. Interbreeding has suggested that the transgene has disrupted a gene allelic with the hotfoot locus located on chromosome 6.

Purkinje cell degeneration* (pcd) *locus The transgenic mouse strain *p432* was produced after zygote microinjection and contains several copies of an SV40-DHFR transgene (Krulewski et al. 1989). Mice homozygous for the transgene exhibit a severe motor disorder characterized by loss of cerebellar Purkinje cells, retinal photoreceptor cells, and olfactory bulb mitral cells and impairment of spermatogenesis. Interbreeding has suggested that the transgene has disrupted a gene allelic with the Purkinje cell degeneration locus located on chromosome 13.

Distonia* (dt) *locus The transgenic mouse strain *Tg4* was generated by pronuclear injection and contains 15–20 copies of an *hsp68-lacZ* transgene (Kothary et al. 1988). Newborn mice homozygous at the transgene locus display increasingly uncoordinated limb movements, and death occurs shortly after weaning. Histological examination reveals a severe loss of dorsal spinal root sensory axons. The *hsp68-lacZ* transgene is expressed only in the medial ventral edge of the neural tube in midgestation embryos. An interesting possibility is that the pattern of transgene expression reflects the expression of the disrupted gene. The phenotype of the *Tg4* insertional mutant is identical to the phenotype of the spontaneous mouse mutation dystonia muscularium, and interbreeding has suggested that these mutations are allelic.

Hyperplastic egg cylinder (hec) **phenotype** The transgenic mouse strain *413* was derived after ES cells were infected with a recombinant retrovirus (M. Kuehn et al., pers. comm.). Mice homozygous for the presence of the acquired provirus fail to develop past the early postimplantation stage and die prior to the formation of the primitive streak. Embryonic death is preceded by hyperplasia of all cell types except the mesoderm, which fails to form properly.

SUMMARY AND CONCLUSIONS

The number of mouse strains carrying insertional mutations has increased steadily since the introduction of transgenic techniques in the early 1980s. The majority of insertional mutants have been recovered serendipitously after producing transgenic mice for some other purpose. Insertional mutations have been recovered at similar frequencies after retroviral infection or microinjection of cloned DNA into zygotes. However, mutations produced by retroviral infection have been easier to analyze than those produced by DNA microinjection. Retroviral integration does not produce gross rearrangements, and thus the molecular analysis of most retrovirally induced insertional mutations has been possible. To date, cDNAs have been cloned in six of eight insertional mutations caused by newly acquired proviruses. However, potential proviral integration sites may represent a subset of all mutable genes, as proviruses seem to preferentially integrate into transcriptionally active chromatin.

In contrast, microinjection of zygotes with cloned DNA can result in severe rearrangements or deletions of chromosomal sequences around an integration site containing many tandem copies of the transgene. In only one of these cases has cDNA cloning of the transcript from the mutated gene been reported, and it is likely that only a fraction of insertional mutations induced by DNA microinjection are amenable to molecular analysis. It is not known whether microinjected DNA integrates randomly into the genome or integrates preferentially into a subset of chromatin.

The formal genetic proof that exhibition of a mutant phenotype is caused by a disruption of the candidate gene has yet to be provided for any of the insertional mutants we have described. The identity of the mutated gene should be confirmed by introduction of the wild-type gene into the mutant mouse with subsequent reversion to the normal phenotype. The only example of a successful rescue experiment in the mouse is the reversion of the β_2 microglobulin (β_2-m) null phenotype (Zijlstra et al. 1989, 1990) to wild type after interbreeding with a transgenic strain containing the human β_2-m gene (M. Zijlstra and R. Jaenisch, unpubl.). Genetic rescue is necessary because the phenotype may be a consequence of a mutation in another gene closely linked to the insertion site. Our attempts to rescue the *Mov13* phenotype by interbreeding with a transgenic strain containing a human *Col1A1* gene have only been partially successful (Wu et al. 1990). This result suggests that either the human gene does not function correctly in the mouse or the disruption in type I collagen in *Mov13* mice may not be the sole mutagenic lesion.

The recent introduction of the ES cell system as an alternative method to generate transgenic mice has allowed the development of novel stategies to mutate the mouse insertionally. These strategies make use of vectors that enrich mutagenic integration events because they are designed to function as gene traps. This approach is promising because it permits a more rational and efficient way to isolate and identify insertional mutations, as the ES cell clones containing these vectors can be prescreened for gene disruption before producing an animal. Prescreening can be either direct, via selection of an antibiotic resistance gene, or indirect by detection of the expression of a *lacZ* cassette. Histological staining of *lacZ* activity, transcribed from the promoter of the interrupted gene, should allow prediction of the tissue-specific pattern of expression of the interrupted gene and may provide information as to which tissues will be affected in the homozygous mutant mouse.

The production and analysis of insertional mutations in the mouse allow the discovery of new genes involved in complex biological processes in mammals. This opportunity, to uncover novel biological paradigms, is the primary intellectual attraction of this endeavor. If this approach is to be exploited fully, procedures for the systematic production of insertional mutations must be improved further. At this time, the use of the ES cell system and retroviral gene-trap vectors seems to represent the most promising avenue of research. Hopefully, in the future, improved strategies will allow wider application of this approach, both for random mutagenesis and for generation of mutants affecting specific biological processes in the mouse.

Acknowledgments

We thank Drs. R.L. Brinster, F. Costantini, N. Jenkins, M. Kuehn, P. Overbeek, R. Palmiter, S.S. Potter, U. Ruther, D. Solter, and P. Soriano for communication of unpublished results and J. Kreidberg and K.-F. Lee for critical review of the manuscript. The work from the laboratory of R.J. was supported by National Institutes of Health grant HD-19105 and National Cancer Institute grant PO1-HL 41484. M.A.R. is a postdoctoral fellow of the National Cancer Institute of Canada.

References

Barklis, E., R.C. Mulligan, and R. Jaenisch. 1986. Chromosomal position of virus permits retrovirus expression in embryonal carcinoma cells. *Cell* 47: 391.

Beier, D.R., C.C. Morton, A. Leder, R. Wallace, P. Leder. 1989. Perinatal lethality (ple): A mutation caused by integration of a transgene into distal mouse chromosome 15. *Genomics* 4: 498.

Berger, S.A., and A. Bernstein. 1985. Characterization of a retrovirus shuttle vector capable of either proviral integration or extrachromosomal replication in mouse cells. *Mol. Cell. Biol.* 5: 305.

Bingham, P.M., and Z. Zachar. 1989. Retrotransposons and the FB transposon from *Drosophila melanogaster*. In *Mobile DNA* (ed. D. Berg and M. Howe), p. 485. American Society for Microbiology, Washington, D.C.

Boeke, J.D. 1989. Transposable elements in *Saccharomyces cerevisiae*. In *Mobile DNA* (ed. D. Berg and M. Howe), p. 335. American Society for Microbiology, Washington, D.C.

Bradley, A., M. Evans, M.H. Kaufman, and E. Robertson. 1984. Formation of germ line chimaeras from embryo-derived teratocarcinomas. *Nature* 309: 255.

Breindl, M., K. Harbers, and R. Jaenisch. 1984. Retrovirus-induced lethal mutation in collagen I gene of mice is associated with an altered chromatin structure. *Cell* 38: 9.

Brinster, R.L., H.Y. Chen, M. Trumbauer, A.W. Senear, R. Warren, and R.D. Palmiter. 1981. Somatic expression of herpes thymidine kinase in mice following injection of a fusion gene into eggs. *Cell* 27: 223.

Capecchi, M.R. 1989. The new mouse genetics: Altering the genome by gene targeting. *Trends Genet.* 5: 70.

Cepko, C.L., B.E. Roberts, and R.C. Mulligan. 1984. Construction and applications of a highly transmissible murine retrovirus shuttle vector. *Cell* 37: 1053.

Costantini, F. and E. Lacy. 1981. Introduction of a rabbit β-globin gene into the mouse germ line. *Nature* 294: 92.

Dente, L., U. Ruther, M. Tripodi, E.F. Wagner, and R. Cortese. 1988. Expression of human α 1-acid glycoprotein genes in cultured cells and in transgenic mice. *Genes Dev.* 2: 259.

Dunnett, S.B., D.J. Sirinathsinghji, R. Heavens, D.C. Rogers, and M.R. Kuehn. 1989. Monoamine deficiency in transgenic (HPRT$^-$) mouse model of Lesch-Nyhan syndrome. *Brain Res.* 501: 401.

Erlich, H.A., ed. 1989. *PCR Technology*. Stockton, New York.

Evans, M.J., and M.H. Kaufman. 1981. Pluripotential cells grown directly from normal mouse embryos. *Nature* 292: 154.

Federoff, N. 1983. Controlling elements in maize. In *Mobile genetic elements* (ed. J.A. Shapiro). p. 1. Academic Press, New York.

Finger, S., R.P. Heavens, D.J. Sirinathsinghji, M.R. Kuehn, and S.B. Dunnett. 1988. Behavioral and neurological evaluation of a transgenic mouse model of Lesch-Nyhan syndrome. *J. Neurol. Sci.* 86: 203.

Gordon, J.W. and F.H. Ruddle. 1981. Integration and stable germ line transmission of genes injected into mouse pronuclei. *Science* 214: 1244.

Gordon, J.W., G.A. Scangos, D.J. Plotkin, J.A. Barbosa, and F.H. Ruddle. 1980.

Genetic transformation of mouse embryos by microinjection of purified DNA. *Proc. Natl. Acad. Sci.* **77:** 7380.
Gordon, J.W., J. Uehlinger, N. Dayani, B.E. Talansky, M. Gordon, G.S. Rudomen, and P.E. Neumann. 1990. Analysis of the hotfoot (ho) locus by creation of an insertional mutation in a transgenic mouse. *Dev. Biol.* **137:** 349.
Gossler, A., A.L. Joyner, J. Rossant, and W. Skarnes. 1989. Mouse embryonic stem cells and reporter constructs to detect developmentally regulated genes. *Science* **244:** 463.
Gossler, A., T. Doetschman, R. Korn, E. Serfling, and R. Kemler. 1986. Transgenesis by means of blastocyst-derived embryonic stem cell lines. *Proc. Natl. Acad. Sci.* **83:** 9065.
Gridley, T., D.A. Gray, T. Orr-Weaver, P. Soriano, D.E. Barton, U. Francke, and R. Jaenisch. 1990. Molecular analysis of the Mov34 mutation: Transcript disrupted by proviral integration in mice is conserved in *Drosophila*. *Development* **109:** 235.
Gridley, T., P. Soriano, and R. Jaenisch. 1987. Insertional mutagenesis in mice. *Trends Genet.* **3:** 162.
Hanahan, D. 1989. Transgenic mice as probes into complex systems. *Science* **246:** 1265.
Harbers, K., M. Kuehn, H. Delius, and R. Jaenisch. 1984. Insertion of retrovirus into the first intron of α1(I) collagen gene leads to embryonic lethal mutation in mice. *Proc. Natl. Acad. Sci.* **81:** 1504.
Hartung, S., R. Jaenisch, and M. Breindl. 1986. Retrovirus insertion inactivates mouse α 1(I) collagen gene by blocking initiation of transcription. *Nature* **320:** 365.
Hogan, B., F. Costantini, and E. Lacy. 1986. *Manipulating the mouse embryo. A laboratory manual.* Cold Spring Harbor Laboratory, Cold Spring Harbor, New York.
Jaenisch, R. 1976. Germ line integration and Mendelian transmission of the exogenous Moloney leukemia virus. *Proc. Natl. Acad. Sci.* **73:** 1260.
———. 1977. Germ line integration of Moloney leukemia virus: Effect of homozygosity at the M-MuLV locus. *Cell* **12:** 691.
———. 1980. Retroviruses and embryogenesis: Microinjection of Moloney leukemia virus into midgestation mouse embryos. *Cell* **19:** 181.
———. 1988. Transgenic animals. *Science* **240:** 1468.
Jaenisch, R. and B. Mintz. 1974. Simian virus 40 DNA sequences in DNA of healthy adult mice derived from preimplantation blastocysts injected with viral DNA. *Proc. Natl. Acad. Sci.* **71:** 1254.
Jaenisch, R., D. Jahner, P. Nobis, I. Simon, J. Lohler, K. Harbers, and D. Grotkopp. 1981. Chromosomal position and activation of retroviral genomes inserted into the germ line of mice. *Cell* **24:** 519.
Jaenisch, R., K. Harbers, A. Schnieke, J. Lohler, I. Chumakov, D. Jahner, D. Grotkopp, and E. Hoffmann. 1983. Germline integration of Moloney leukemia virus at the Mov 13 locus leads to recessive lethal mutation and early embryonic death. *Cell* **32:** 209.
Jahner, D. and R. Jaenisch. 1985. Retrovirus-induced de novo methylation of flanking host sequences correlates with gene inactivity. *Nature* **315:** 594.
Jenkins, N, and N. Copeland. 1985. High frequency acquisition of ecotropic MuLV proviruses in SWR/J-RF/J hybrid mice. *Cell* **43:** 811.

Jenkins, N., N. Copeland, B.A. Taylor, and B.K. Lee. 1981. Dilute (d) coat colour mutation of DBA/2J mice is associated with the site of integration of an ecotropic MuLV genome. *Nature* **293:** 370.

Keller, S.A., S. Liptay, A. Hajra, and M.H. Meisler. 1990. Transgene-induced mutation of the murine steel locus. *Proc. Natl. Acad. Sci.* **87:** 10019.

Kotheray, R., S. Clapoff, A. Brown, R. Campbell, A. Peterson, and J. Rossant. 1988. A transgene containing lacZ inserted into the dystonia locus is expressed in neural tube. *Nature* **335:** 435.

Kratochwil, K., K. von der Mark, E.J. Koller, R. Jaenisch, K. Mooslehner, M. Schwartz, K. Haase, I. Gmachl, and K. Harbers. 1989. Retrovirus-induced mutation in Mov13 affects collagen I expression in a tissue-specific manner. *Cell* **57:** 807.

Krulewski, T.F., P.E. Neumann, and J.W. Gordon. 1989. Insertional mutation in a transgenic mouse allelic with Purkinje cell degeneration. *Proc. Natl. Acad. Sci.* **86:** 3709.

Kuehn, M.R., A. Bradley, E.J. Robertson, and M.J. Evans. 1987. A potential animal model for Lesch-Nyhan syndrome through induction of HPRT mutations in mice. *Nature* **326:** 295.

Lacy, E., S. Roberts, E.P. Evans, M.D. Burtenshaw, and F. Costantini. 1983. A foreign β-globin gene in transgenic mice: Integration at abnormal chromosomal positions and expression in inappropriate tissues. *Cell* **34:** 343.

Lohler, J., R. Timpl, and R. Jaenisch. 1984. Embryonic lethal mutation in mouse collagen I gene causes rupture of blood vessels and is associated with erythropoietic and mesenchymal cell death. *Cell* **38:** 597.

Maas, R.L., R. Zeller, R.P. Woychik, T.F. Vogt, and P. Leder. 1990. Disruption of formin-encoding transcripts in two mutant limb deformity alleles. *Nature* **346:** 853.

MacGregor, G.R., L.D. Russell, M.E.A.B. van Beek, G.R. Hanten, M.J. Kovac, C.A. Kozak, M.L. Meistrich, and P.A. Overbeek. 1990. Symplastic spermatids (sys): A recessive insertional mutation in mice causing a defect in spermatogenesis. *Proc. Natl. Acad. Sci.* **87:** 5016.

Mann, R., R.C. Mulligan, and D. Baltimore. 1983. Construction of a retrovirus packaging mutant and its use to produce helper-free defective retrovirus. *Cell* **33:** 153.

Martin, G.R. 1981. Isolation of a pluripotential cell line from early mouse embryos cultured in medium conditioned by teratocarcinoma stem cells. *Proc. Natl. Acad. Sci.* **78:** 7634.

McNeish, J.D., W.J. Scott, and S.S. Potter. 1988. Legless, a novel mutation found in pHT1-1 transgenic mice. *Science* **241:** 837.

McNeish, J.D., J. Thayer, K. Walling, K.S. Sulik, S.S. Potter, and W.J. Scott. 1990. Phenotypic characterization of the transgenic mouse insertional mutation, legless. *J. Exp. Zool.* **253:** 151.

Mercer, J.A., P.K. Seperack, M.C. Strobel, N.G. Copeland, and N.A. Jenkins. 1991. The murine dilute coat color locus encodes a novel myosin heavy chain. *Nature* **349:** 709.

Overbeek, P.A., S.-P. Lai, K.R. van Quill, and H. Westphal. 1986. Tissue-specific expression in transgenic mice of a fused gene containing RSV terminal sequences. *Science* **231:** 1574.

Palmiter, R.D. and R.L. Brinster. 1986. Germline transformation of mice. *Annu.*

Rev. Genet. **20:** 465.

Peckham, I., S. Sobel, J. Comer, R. Jaenisch, and E. Barklis. 1989. Retrovirus activation in embryonal carcinoma cells by cellular promoters. *Genes Dev.* **3:** 2062.

Pohl, T.M., M.-G. Mattei, and U. Ruther. 1990. Evidence for allelism of the recessive insertional mutant add and the dominant mouse mutation extra-toes (Xt). *Development* **110:** 1153.

Radice, G., J.J. Lee, and F. Costantini. 1991. HB58, an insertional mutation affecting early post-implantation development of the mouse embryo. *Development* **111:** 801.

Reik, W., H. Weiher, and R. Jaenisch. 1985. Replication competent Moloney leukemia virus carrying a bacterial suppressor tRNA gene: Selective cloning of proviral and flanking host sequences. *Proc. Natl. Acad. Sci.* **82:** 1141.

Rinchik, E.M., L.B. Russell, N.G. Copeland, and N.A. Jenkins. 1985. The dilute-short (d-se) ear complex of the mouse: Lessons from a fancy mutation. *Trends Genet.* **1:** 170.

Robertson, E.J., ed. 1987. *Teratocarcinomas and embryonic stem cells: A practical approach.* IRL Press, Oxford.

Robertson, E.J., A. Bradley, M. Keuhn, and M. Evans. 1986. Germ-line transmission of genes introduced into cultured pluripotential cells by retroviral vector. *Nature* **323:** 445.

Roeder, G.S. and G.R. Fink. 1980. DNA rearrangements associated with a transposable element in yeast. *Cell* **21:** 239.

Rohdewohld, H., H. Weiher, W. Reik, R. Jaenisch, and M. Breindl. 1987. Retrovirus integration and chromatin structure: Moloney leukemia proviral integration sites map near DNase I hypersensitive sites. *J. Virol.* **61:** 336.

Rossant, J. and A.L. Joyner. 1989. Towards a molecular genetic analysis of mammalian development. *Trends Genet.* **5:** 277.

Rudnicki, M.A. and M. McBurney. 1987. Cell culture methods and induction of differentiation of embryonal carcinoma cell lines. In *Teratomas and embryonic stem cells: A practical approach* (ed. E.J. Robertson), chap. 2. IRL Press, Oxford.

Schnieke, A., K. Harbers, and R. Jaenisch. 1983. Embryonic lethal mutation in mice produced by retroviral insertion into the alpha 1 (I) collagen gene. *Nature* **304:** 315.

Seliger, B., R. Kollek, C. Stocking, T. Franz, and W. Ostertag. 1986. Viral transfer, transcription, and rescue of a selectable myeloproliferative sarcoma virus in embryonal cell lines: Expression of the mos oncogene. *Mol. Cell. Biol.* **6:** 286.

Shani, M. 1986. Tissue-specific and developmentally regulated expression of a chimeric actin-globin gene in transgenic mice. *Mol. Cell. Biol.* **6:** 2624.

Shawlot, W., M.J. Siciliano, R.L. Stallings, and P.A. Overbeek. 1989. Insertional inactivation of the *downless* gene in a family of transgenic mice. *Mol. Biol. Med.* **6:** 299.

Shih, C.-C., J.P. Stoye, and J.M. Coffin. 1988. Highly preferred targets for retrovirus integration. *Cell* **53:** 531.

Silver, J. and V. Keerikatte. 1989. Novel use of polymerase chain reaction to amplify cellular DNA adjacent to an integrated provirus. *J. Virol.* **63:** 1924.

Soriano, P. and R. Jaenisch. 1986. Retroviruses as probes for mammalian development: Allocation of cells to the somatic and germ cell lineages. *Cell* **46:** 19.

Soriano, P., T. Gridley, and R. Jaenisch. 1987. Retroviruses and insertional mutagenesis in mice: Proviral intagration at the Mov 34 locus leads to early embryonic death. *Genes Dev.* **1:** 366.

Spence, S.E., D.J. Gilbert, D.A. Swing, N.G. Copeland, and N.A. Jenkins. 1989. Spontaneous germ line virus infection and retroviral insertional mutagenesis in eighteen transgenic *Srev* lines of mice. *Mol. Cell. Biol.* **9:** 177.

Spradling, A.C. and G.M. Rubin. 1981. *Drosophila* genome organization: Conserved and dynamic aspects. *Annu. Rev. Genet.* **20:** 219.

Stewart, T.A., P.K. Pattengale, and P. Leder. 1984. Spontaneous mammary adenocarcinomas in transgenic mice that carry and express MTV/myc fusion genes. *Cell* **38:** 627.

Stoye, J.P., S. Fenner, G.E. Greenoak, C. Moran, and J.M. Coffin. 1988. Role of endogenous retroviruses as mutagens: The hairless mutation of mice. *Cell* **54:** 383.

Strobel, M.C., P.K. Seperack, N.G. Copeland, and N.A. Jenkins. 1990. Molecular analysis of two mouse dilute locus deletion mutations: Spontaneous dilute lethal 20J and radiation-induced dilute prenatal lethal Aa2 alleles. *Mol. Cell. Biol.* **10:** 501.

Stuhlmann, H., R. Cone, R.C. Mulligan, and R. Jaenisch. 1984. Introduction of a selectable gene into different animal tissue by a retroviral recombinant vector. *Proc. Natl. Acad. Sci.* **81:** 7151.

Surani, M.A., W. Reik, and N.D. Allen. 1988. Transgenes as molecular probe for genomic imprinting. *Trends Genet.* **4:** 59.

Thomas, K.R. and M.R. Capecchi. 1987. Site directed mutagenesis by gene targeting in mouse embryo-derived stem cells. *Cell* **51:** 503.

Varmus, H. 1988. Retroviruses. *Science* **240:** 1427.

Varmus, H. and P. Brown. 1989. Retroviruses. In *Mobile DNA* (ed. D. Berg and M. Howe), p. 53. American Society for Microbiology, Washington, D.C.

Vijaya, S., D.L. Steffen, and H.L. Robinson. 1986. Acceptor sites for retroviral integration map near DNase I hypersensitive sites in chromatin. *J. Virol.* **60:** 683.

von Melchner, H. and E.H. Ruley. 1989. Identification of cellular promoters by using a retrovirus promoter trap. *J. Virol.* **63:** 3227.

von Melchner, H., S. Reddy, and H.E. Ruley. 1990. Isolation of cellular promoters by using a retrovirus promoter trap. *Proc. Natl. Acad. Sci.* **87:** 3733.

Wagner, E.F., T.A. Stewart, and B. Mintz. 1981. The human β-globin gene and a functional thymidine kinase gene in developing mice. *Proc. Natl. Acad. Sci.* **78:** 5016.

Wagner, E.F., L. Covarriubias, T.A. Stewart, and B. Mintz. 1983. Prenatal lethalities in mice homozygous for human growth hormone gene sequences integrated in the germ line. *Cell* **35:** 647.

Wagner, T.E., P.C. Hoppe, J.D. Jollick, D.R. Scholl, R.L. Hodinka, and J.B. Gault. 1981. Microinjection of a rabbit β-globin into zygotes and its subsequent expression in adult mice and their offspring. *Proc. Natl. Acad. Sci.* **78:** 6376.

Weiher, H., E. Barklis, W. Ostertag, and R. Jaenisch. 1987. Two distinct sequence elements mediate retroviral gene expression in embryonal carcinoma cells. *J. Virol.* **62:** 2742.

Weiher, H., T. Noda, D.A. Gray, A.H. Sharpe, and R. Jaenisch. 1990. Transgenic mouse model of kidney disease: Insertional inactivation of ubiquitously expressed gene leads to nephrotic syndrome. *Cell* **62:** 425.

Weiss, R., N. Teich, H. Varmus, and J. Coffin, eds. 1984. *RNA tumor viruses.* Cold Spring Harbor Laboratory, Cold Spring Harbor, New York.

Woychik, R.P., R.L. Maas, R. Zeller, T.F. Vogt, and P. Leder. 1990. "Formins": Proteins deduced from the alternative transcripts of the limb deformity gene. *Nature* **346:** 850.

Woychik, R.P., T.A. Stewart, L.G. Davis, P. D'Estachio, and P. Leder. 1985. An inherited limb deformity created by insertional mutagenesis in a transgenic mouse. *Nature* **318:** 36.

Wu, H., J.F. Bateman, A. Schnieke, A. Sharpe, D. Barker, T. Mascara, D. Eyre, R. Bruns, P. Krimpenfort, A. Burns, and R. Jaenisch. 1990. Human-mouse interspecies collagen I heterodimer is functional during embryonic development of Mov13 mutant mouse embryos. *Mol. Cell. Biol.* **10:** 1452.

Xiang, X., K.F. Benson, and K. Chada. 1990. Mini-mouse: Disruption of the pygmy locus in a transgenic insertional mutant. *Science* **247:** 967.

Zeller, R., L. Jackson-Grusby, and P. Leder. 1990. The limb deformity gene is required for apical ectodermal ridge differentiation and anteroposterior limb pattern formation. *Genes Dev.* **3:** 1481.

Zijlstra, E. Li, F. Sajjadi, S. Subramani, and R. Jaenisch. 1989. Germ-line transmission of a disrupted β2-microglobulin gene produced by homologous recombination in embryonic stem cells. *Nature* **342:** 435.

Zijlstra, M., M. Bix, N.E. Simister, J.M. Loring, D.H. Raulet, and R. Jaenisch. 1990. β2-microglobulin mice lack $CD4^-8^+$ cytolytic T cells. *Nature* **344:** 742.

Chromosome Imprinting and Its Significance for Mammalian Development

Bruce M. Cattanach

MRC Radiobiology Unit
Chilton
Oxon OX11 0RD, England

Since the rediscovery of Mendel's laws of inheritance at the turn of the century, it has been a basic tenet of genetics that nuclear genes inherited from each parent function equally in the offspring. Countless breeding experiments with diverse plant and animal species, using a wide variety of genetic traits, have testified to the general validity of the theory. Nevertheless, over the past 50 years, evidence has been accumulating that Mendel's laws are not always correct. The earliest exceptions were regarded as bizarre, novel events of little general significance. However, following the recent discovery that both maternal and paternal genomes are necessary for normal mammalian embryonic development and, more specifically, that genes in certain chromosome regions function differently according to parental origin (Solter 1988; Hall 1990), interest in the subject has developed rapidly. The concept has developed that the maternal and paternal copies are differentially marked, or imprinted, in the parental germ lines, with the consequence that they become functionally inequivalent in the embryo.

This chapter reviews the evidence on imprinting effects in mammals, notably mice and humans, and discusses the role of imprinting in mammalian development.

Specific topics addressed are:

❏ the need for both maternal and paternal genomes in development

- the need for both maternal and paternal copies of certain chromosome regions and genes in development
- indications of imprinting in humans
- imprinting of transgenes in mice
- imprinting of the X chromosome
- the role of imprinting

INTRODUCTION

It has long been known that chromosomes of maternal and paternal origin behave differently in the fly, *Sciara*, and in mealy bug, coccid, species (Metz 1938; Hughes-Schrader 1948). However, it was Crouse (1960) who first introduced the term imprinting to describe the process that leads to the complex series of selective eliminations of the paternal X in the soma and germ lines of *Sciara coprophila*, which is seen as the basis of sex determination in this species. Thus, the maternal and paternal chromosomes appear to be marked in some way that allows them to be identified through gametogenesis in the progeny. At this point, the original imprint must be lost and a new imprint gained. An interesting point is that the X chromosome behavior in *Sciara* is dependent on a single controlling locus that lies close to the centromere.

In coccids, a differential behavior of the maternally and paternally derived chromosomes is also the basis of sex determination, but here, heterochromatinization and genetic inactivity replace chromosome loss (although chromosome loss ultimately occurs in one coccid system), and the whole paternal chromosome set is imprinted to exhibit this behavior (Nur 1990).

In more recent times, the term imprinting has been used to describe the parental chromosome strand-dependent switching of mating type in yeast (Klar 1990) and the differences in activities of maternally and paternally derived genomes in maize endosperm (Kermicle and Alleman 1990).

The phenomenon of imprinting is now widely recognized in mammals, notably in mice and humans, where it appears to play an important role in embryonic development (Solter 1988). Imprinting also appears to be responsible for irregular patterns of inheritance and variable expressions of a number of human disorders and in establishing events predisposing to certain sporadic forms of cancer (Hall 1990). These aspects of imprinting form the core of this review.

THE NEED FOR BOTH MATERNAL AND PATERNAL GENOMES IN DEVELOPMENT

Pronuclear transplantation experiments

A first indication that maternal and paternal genomes are required for mammalian embryonic development comes from the observation that successful parthenogenesis does not occur (Solter 1988). In the mouse, for example, parthenogenotes can complete preimplantation development and initiate implantation, but they fail to go to term. Development usually terminates at midgestation (Kaufman 1983). The absence of some extragenomic contribution from the sperm and homozygosity for lethal genes were the two main hypotheses offered to explain the failure of parthenogenetic embryos, but mouse pronuclear transplantation experiments in which male and female pronuclei can be transferred between embryos negated such explanations and instead demonstrated the need for both maternal and paternal genomes (Solter 1988).

The lack of some cytoplasmic factor contributed by the sperm was excluded as the cause of parthenogenote failure when it was found that uniparental gynogenetic embryos, derived by removal of the male pronucleus following suppression of the second polar body extrusion, failed to survive (Surani and Barton 1983). Likewise, embryos produced from diploid parthenogenetic eggs, following transfer of their pronuclei into enucleated fertilized eggs, also died soon after fertilization (Mann and Lovell-Badge 1984).

Homozygosity for lethal genes was excluded when it was found that gynogenetic embryos, produced by transferring female pronuclei into haploid fertilized eggs, showed the same developmental deficiencies as parthenogenotes (McGrath and Solter 1984a,b). Their diparental origin and effectively greater heterozygosity did not offer any advantage to their development. In contrast, these studies and those of Surani et al. (1984) and Barton et al. (1984) demonstrated the absolute need for both maternal and paternal genomes. Thus, whereas such gynogenetic embryos showed poor postimplantation development, the transfer of a male pronucleus into a haploid parthenogenetic egg allowed development to term. Similarly, exchanges of male and female pronuclei from fertilized eggs showed that failure of development occurred with both gynogenetic (two female pronuclei) and androgenetic (two male pronuclei) eggs, whereas reconstituted normal eggs (one male plus one female pronucleus) could survive normally.

A key observation in these studies was that embryos with two female pronuclei developed relatively normally up to day 10 of gestation but failed in trophoblast development, whereas embryos with two male pronuclei developed more poorly but showed an overgrowth of the trophoblast. It was concluded from these findings that as a result of epigenetic modifications during parental gametogenesis, the maternal and paternal genomes were differentially imprinted, with the consequence that they were not functionally equivalent in the embryo (Barton et al. 1984; Surani et al. 1984).

Later studies demonstrated that the imprint persisted throughout the early cleavage stages (Surani et al. 1986a,b). Thus, transplantation of nuclei from either 2-4-celled androgenotes or 2-8-celled parthenogenotes into fertilized eggs from which one pronucleus had been removed could allow development to term, but only when the single pronucleus in the egg was of opposite parental origin. The functional difference between the parental genomes was therefore heritable through several cleavage divisions and survived activation and reprogramming of the embryonic genome.

Reconstruction experiments with chimeras

The opposite natures of parthenogenetic and androgenetic embryo development suggested that the maternal genome may be more important for the development of the embryo proper, whereas the paternal genome may be important for the development of the trophoblast. Reconstruction experiments have indicated that the situation is more complex, however. In one study (Barton et al. 1985), inner cell masses from parthenogenetic or gynogenetic blastocysts were introduced into trophectoderm vesicles from normal blastocysts, and although subsequent development of the embryos was improved, they still failed to survive to term. Clearly, a paternal genome was required for normal development of the embryo as well as the trophoblast (Surani et al. 1986a). A similar conclusion could also be drawn from a further study (Surani et al. 1987) in which chimeras were made from parthenogenetic-androgenetic combinations. Even though both parental genomes were present, if in different cells, there was incomplete functional complementation as the chimeras did not go to term. A striking finding in these studies was that the parthenogenetic cells became almost entirely confined to the embryo and the androgenetic cells became confined to the trophoblast, with the yolk sac containing cells of both types. Similar findings have been made with parthenogenetic-normal chimeras (Clarke et al. 1988). A spatial specificity of the two cell types was therefore indicated.

Rescue experiments have also demonstrated the need for both maternal and paternal genomes in various differentiated tissues at later stages of development. Parthenogenetic cells in chimeras with normal cells can contribute to a variety of cell lineages, especially the germ line of females (Anderegg and Markert 1986), but there is strong selection against them after day 13 of gestation in most tissues, notably in skeletal muscle, pancreas, and tongue (Nagy et al. 1987; Surani et al. 1988b; Fundele et al. 1989, 1990). In contrast, androgenetic cells in chimeras appear to be relatively immune to cell selection up to day 15 of gestation, they contribute more to muscle, skeleton, and heart than do parthenogenetic cells, and they are associated with an increased size of the conceptus (Surani et al. 1990a). A large contribution is also associated with phenotypic abnormalities of skeletal elements (Surani et al. 1990b). Consistent with these observations is the finding that embryonic stem (ES) cells derived from androgenetic blastocysts gave rise to tumors composed almost entirely of striated muscle when transplanted to ectopic subcutaneous sites; parthenogenote-derived stem cells, on the other hand, gave rise to tumors consisting of a variety of tissues (Mann et al. 1990). These studies also showed that chimeras derived from the introduction of androgenetic ES cells into normal blastocysts developed skeletal anomalies, particularly in the ribs.

Conclusions

Observations of early embryos and chimeras allow a number of important conclusions to be drawn regarding imprinting and its effects. Imprinting effects can occur very early in development and persist through to term and possibly thereafter. Effects on growth are consistently found, as in the early embryo of androgenotes and the trophoblast of parthenogenotes. Chimera studies further illustrate the abilities and inabilities of cells with only maternal or paternal genomes to contribute to various cell lineages. Opposite effects are quite consistently found. A problem for interpretation of these data is that they represent the cumulative effects of the presence of whole maternal or paternal genomes. The possibility that they represent the consequences of interactions of specific genes subject to different imprinting effects cannot be evaluated. It is important to note that none of these effects have been found to be influenced by genetic background (Surani et al. 1990b).

THE NEED FOR BOTH MATERNAL AND PATERNAL COPIES OF CERTAIN CHROMOSOME REGIONS AND GENES IN DEVELOPMENT

The first evidence that maternal and paternal copies of chromosome regions might not be functionally equivalent was obtained by Snell (1946) in studies with a reciprocal translocation in the mouse. Intercrosses between translocation heterozygotes were being used to order marker genes with respect to the centromeres, and, in one test, an expected class did not appear. Similar findings were later obtained using other translocations (Searle et al. 1971; Searle and Beechey 1978; Lyon et al. 1972; Lyon and Glenister 1977). However, it was only after anomalous, opposite phenotypes were found associated with maternal or paternal duplication/deficiencies of proximal regions of chromosome 11 and distal regions of chromosome 2 (Cattanach and Kirk 1985) that chromosome parental origin effects and the reality of chromosome imprinting were generally recognized.

The genetic techniques used to detect these imprinting effects have been described in numerous papers (see, e.g., Cattanach 1986). The basic procedure is one of intercrossing heterozygotes for translocations and, with the use of marker genes, ascertaining from the progeny whether gametes with reciprocal types of chromosomal imbalance can complement each other and give rise to normal, viable zygotes. The simplest system, which is used for investigating whole chromosomes, employs Robertsonian translocations that in heterozygotes cause high frequencies of nondisjunction. By these means, maternal and paternal disomies for any chosen chromosome can be readily generated and investigated.

Equivalent intercrosses using reciprocal translocations are used for detecting regions of individual chromosomes that are subject to imprinting. Maternal duplications with corresponding paternal deficiencies (and vice versa) for a region of a chosen chromosome can be effectively generated, particularly so for the distal regions of chromosomes where the recovery rate is high (16%). One difficulty with this system is that because reciprocal translocations involve two chromosomes, tests with more than one translocation are required to assign any imprinting effect found to a particular chromosome.

In the summary of the results that follow, the term imprinting region is used in the sense that an anomalous phenotype, or imprinting effect, is found when the contributions of maternal and paternal copies are other than normal. As the translocations only crudely define the imprinting regions and these are reduced in size as the investigation proceeds, it is assumed that only one or a very few genes are responsible for the phenomena observed. However, other genes within or close to a region may be subject to the same imprinting effect.

Most of the mouse genome has now been tested using one of the above methods so that only small parts of chromosomes 7, 8, 10, and 18 and all of chromosome 12 remain to be investigated. No effects of either maternal or paternal disomy have been found for chromosomes 3, 4, 13, 15, 16, and 19 (Searle and Beechey 1985; Cattanach 1986; Berger and Epstein 1989; Cattanach and Beechey 1990a,b).

Defined imprinting regions and imprinting phenomena

Imprinting regions and imprinting effects have been identified on five chromosomes. These regions are illustrated in Figure 1.

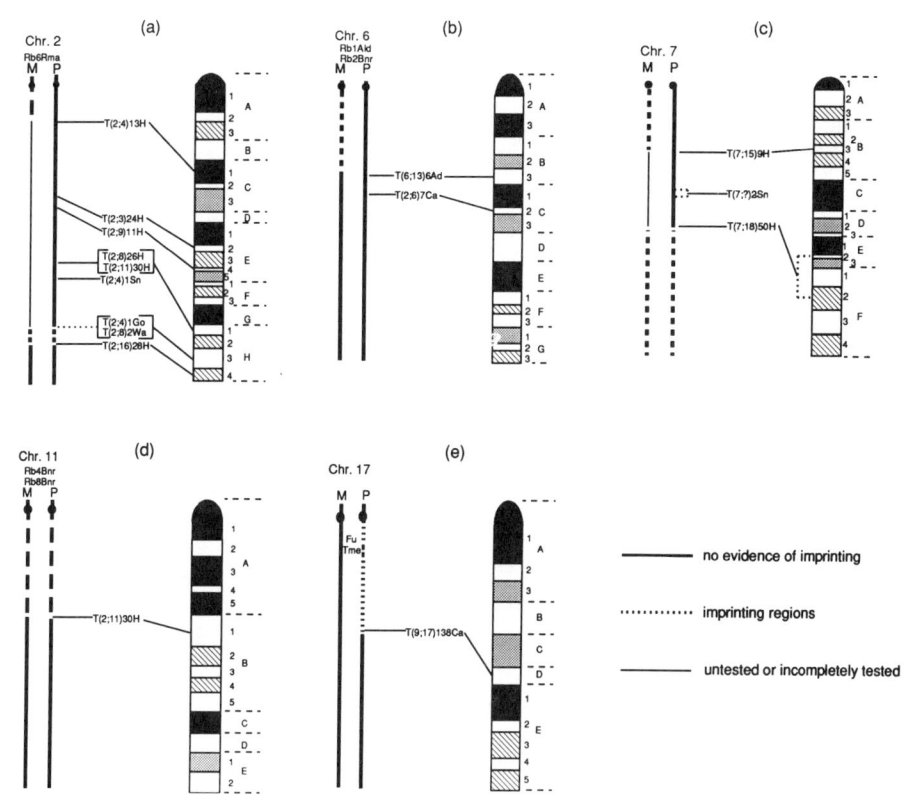

Figure 1 Imprinting regions of mouse autosomal chromosomes. Linkage and physical maps of chromosome 2 (a), chromosome 6 (b), chromosome 7 (c), chromosome 11 (d), and chromosome 17 (e). Positions of translocations used to define the locations of the imprinting regions are illustrated. (M) Maternal chromosomes; (P) paternal chromosomes. Because an imprinting effect that can only be detected after birth has now been found (distal chromosome 17), the central region of chromosome 2 as shown is incompletely tested (Cattanach and Beechey 1990a,b).

Chromosome 2 Two distinct regions of chromosome 2 show imprinting effects. Maternal duplication/paternal deficiency and paternal duplication/maternal deficiency for a distal region have been found to cause different neonatal lethalities with opposite anomalous phenotypes and behavior (Cattanach and Kirk 1985; Cattanach 1986; Cattanach and Beechey 1990b). These two phenotypic classes have been found with each of the chromosome 2 translocations shown in Figure 1a, with the exception of T28H. The findings with T1Go are new (Cattanach et al. 1991) and localize the imprinting region to the small segment of the chromosome between the T1Go and T28H breakpoints in bands 2H3 and 2H4 (Fig. 1a).

The opposite natures of the maternal and paternal duplication deficiency phenotypes clearly suggest that maternal and paternal copies of a gene or genes within the imprinting region function differently at some stage(s) of development; both may be programmed so as to give quantitatively different amounts of some product(s) such that optimum balance is only achieved in the presence of both.

The second chromosome 2 imprinting effect involves the proximal region of the chromosome. Maternal duplication/paternal deficiency for this region gives an early (by day 11 of gestation) embryonic lethality; the equivalent paternal duplication/maternal deficiency genotype is without detectable effect. The lethality was originally detected in studies with T30H, but from work with the other translocations shown in Figure 1a (Cattanach and Kirk 1985; Cattanach and Beechey 1990b), the gene(s) responsible must lie between the T13H breakpoint in band 2C1 and the centromere. The finding that abnormality only occurs with maternal duplication/paternal deficiency suggests that in some early embryonic stage, the maternal copy of a gene within the region is nonfunctional (i.e., suffers imprinting inactivation). Only the paternal copy may be active. An alternative explanation is that the paternal copy is relatively inactive and that two fully active maternal copies cause the lethality, although this possibility is less likely.

Chromosome 6 The imprinting effect found for a proximal region of chromosome 6 resembles that for proximal chromosome 2 in that maternal duplication/paternal deficiency again results in early embryonic lethality prior to day 11 of gestation (Cattanach and Beechey 1990a,b). Studies using the translocations shown in Figure 1b have localized the region responsible for the defect to the area between the T6Ad breakpoint in band 6B3 and the centromere. As suggested for proximal chromosome 2, an imprinting inactivation of the maternal copy of some gene in this region may be indicated.

Chromosome 7 Chromosome 7 is one of the most interesting chromosomes subject to imprinting not only because it may have three, or possibly more, imprinting phenomena associated with it (Searle and Beechey 1990), but also because it is proving to be the most amenable to further investigation.

One imprinting phenomenon is associated with the region proximal to the T9H breakpoint in band 7B3 (Fig. 1c). Maternal duplication/paternal deficiency of this region results in a late onset fetal lethality, with placental and fetal growth being retarded. Maternal duplication/paternal deficiency for a region distal to the T50H breakpoint in band 7F1 provides the second imprinting phenomenon for this chromosome. Lethality is again the consequence, with retardation of fetal and placental growth and, ultimately, fetal death occurring earlier (at about midgestation). Paternal duplication/maternal deficiency of the same distal region gives the third imprinting effect, and this comprises an early embryonic lethality that occurs before day 11 of gestation. It may be noted that the region between the T9H and T50H breakpoints has not been studied effectively in the paternal copy or at all in the maternal copy for imprinting effects that occur postnatally. However, on the basis of homology with a segment of a human chromosome that has shown seemingly good imprinting phenomena (see below), it is possible that this region of mouse chromosome 7 may also be subject to imprinting.

Further evidence on the imprinting in the distal region of chromosome 7 has come from two other lines of investigation. Collaborative work between Harwell and Cambridge (A.C. Ferguson-Smith et al., in prep.) has successfully rescued the paternal duplication/maternal deficiency genotype in chimeras. The remarkable feature of these chimeras was that they were larger (1.5× normal) than their non-chimeric sibs and had correspondingly large placentas. This again raises the possibility that the imprinting leads to the inactivation of one parental copy of an important growth gene, in this case the maternal copy, such that the possession of two active paternal copies causes the overgrowth. Inactivity of maternal copies in the maternal duplication/paternal deficiency progeny may account for the lethality of this class. The cause of the earlier lethality of the reciprocal paternal duplication/maternal deficiency class, however, remains open. It could be based on an overgrowth characteristic or it could represent an independent imprinting effect involving a different gene.

The second line of investigation pertaining to distal chromosome 7 imprinting relates to the work of DeChiara et al. (1990) with a disrupted gene for insulin-like growth factor (*Igf2*), which is located in this region of chromosome 7 (T. Glaser, unpubl.). The key observation was that heterozygotes receiving the mutated *Igf2* gene from the father were only about 60% of normal size at all postnatal ages, whereas those receiving a maternal copy were normal (E.J. Robertson, pers. comm.). The size difference was also evident at 16–18 days of gestation with placental sizes being correlated. These findings indicate that the *Igf2* gene is subject to imprinting and further makes it the candidate locus for the maternal duplication/paternal deficiency lethality. Indeed, independent work of T.M. DeChiara et al. (pers. comm) and A.C. Ferguson-Smith et al. (in prep.) has shown that the maternally inherited *Igf2* gene is not transcribed and that only the paternal copy is functional. Interestingly, the homozygote for the *Igf2* mutation is viable and, although small, does not show any gross phenotypic abnormality.

Yet to be explained is the discordance between the *Igf2* mutant phenotype and that of the maternal duplication/paternal deficiency class. Neither appears to have a functional paternally derived *Igf2* allele, but the former expresses only small size, whereas the latter causes the prenatal lethality. This might suggest that genes other than *Igf2* in the distal chromosome 7 region may be subject to imprinting inactivation. Nevertheless, it is clear from the data that *Igf2* has a mitogenic role in embryonic growth and, as will be seen, this relates to its receptor locus, *Igf2r*, on chromosome 17.

Chromosome 11 The first imprinting anomaly detected involved chromosome 11 (Cattanach and Kirk 1985) and was characterized by effects on growth. Maternal disomy was found to result in small size (0.7× normal) from birth through to adulthood, whereas paternal disomy caused increased size (1.3× normal). As adults, the small and large disomy mice were normal in all respects except size, and both sexes were fertile. The size differences were not maintained into the next generation, indicating a reprogramming of the genome presumably in the germ lines. The size differences have also been detected prior to birth (15–18 days), correlating with placental sizes. More recent work has shown that the ratios decline at earlier ages and may not exist prior to day 12 (C.V. Beechey et al., in prep.). This is a potentially critical finding as it provides the first evidence of the timing of initiation of an imprinting effect.

As suggested for the distal chromosome 2 imprinting anomalies, the opposite size effects noted here suggest that maternal and paternal copies of a gene are programmed to give quantitatively different amounts of product(s) so that optimum balance is only obtained in the presence of both. The reduction in size observed with maternal disomy 11 has also been obtained with maternal duplication/paternal deficiency for the region proximal to the T30H breakpoint in chromosome 11 (Fig. 1d). This defines the imprinting region in which the responsible gene must lie as proximal to band 11B1.

Chromosome 17 Two regions of chromosome 17 are subject to imprinting. One is located proximally in the chromosome and was detected through the use of the hairpin tail (T^{hp}) deletion (Fig. 1e), which can only be maintained by transmission through males; when it is transmitted through females, the deletion behaves as a lethal (Johnston 1974, 1975). Maternally inherited T^{hp} fetuses are cyanotic, edematous, and display indications of polydactyly. The anomalies of T^{hp} inheritance were originally interpreted to mean that a maternal copy of a gene located in the deleted region is required for normal development. This, of course, implies that the paternal allele is nonfunctional.

Deletion mapping has identified the region responsible for the lethality. It has been named T-maternal effect, *Tme* (Winking and Silver 1984). Barlow et al. (1991) have recently provided evidence that the *Igf2r* locus is located in the *Tme* region. More importantly, these authors have clearly established that, in 15-day embryos, the gene is only expressed when inherited from the mother. Significantly, two other genes, plasminogen (*Plg*) and superoxide dismutase (*Sod-2*), located within the *Tme* region were not so affected. The absence of *Igf2r* expression when inherited from the father therefore suggests imprinting inactivation of the paternal *Igf2r* locus, which is consistent with the T^{hp} inheritance. It seems not without significance that both the *Igf2* ligand on chromosome 7 and its receptor, *Igf2r*, on chromosome 17 are subject to imprinting. There is evidence that they are coordinately expressed during prenatal development in tissues such as muscle (Senior et al. 1990). However, it is not clear why the phenotypes associated with nonexpression of *Igf2* and *Igf2r* should differ.

The second region of imprinting in chromosome 17 lies distal to the T138Ca breakpoint in band 17D (Fig. 1e) and is again expressed as a growth differential. Paternal duplication/maternal deficiency for this region is characterized by small size (70% normal), which is notably evident from day 7. It may be noted that the effect is in the opposite direction from that found with *paternal* disomy for chromosome 11. The size difference appears less clear at birth, but this remains to be verified (Cattanach and Beechey 1990b). As it is unlikely that this size effect will be detectable before birth, the onset of this distal chromosome 17 phenomenon may prove to be the latest to occur of any so far found in the mouse.

Other possible examples of imprinting in the mouse

In general accord with the T^{hp} (*Tme*) phenomena has been Lyon and Glenister's (1977) studies on the region of chromosome 17 proximal to the T138Ca breakpoint (Fig. 1e). These authors noted a consistent shortage of paternal duplication/maternal deficiency young relative to the frequency of recovery of the reciprocal class and suggested that this represented a further manifestation of the T^{hp} phenomenon that can now be attributed to the absence of *Igf2r* expression (Barlow et al. (1991). The validity of this conclusion may be considered established even though once again, the phenotypic effects observed are not identical to those seen with T^{hp}. Nevertheless, it has been proposed that the differential recoveries of maternal and paternal disomies noted for other chromosomes may represent further examples of imprinting (Cattanach and Beechey 1990a). Examination of the literature revealed that for chromosome 1, there was a significant excess recovery of maternal disomy over paternal disomy, and for chromosomes 5, 9, and 14, the opposite was the case (Cattanach and Beechey 1990b). An insignificant excess of maternal over paternal disomy was also seen for chromosome 4. A problem with these data, derived from intercrosses using Robertsonian translocations, is the low frequency with which the disomies are generated (Tease and Cattanach 1986).

Repeating the tests using reciprocal translocations to screen for differential recoveries of the maternal and paternal duplication/deficiency classes for the distal regions has so far failed to support the Robertsonian translocation results (Cattanach and Beechey 1990b), although some limited older data are consistent with the chromosome 1 findings (Searle et al. 1971; C.V. Beechey, unpubl.). Studies with the proximal regions will be needed to establish whether the differential recoveries noted in the literature have any significance for imprinting and, consequently, reflect the occurrence of prenatal or neonatal lethalities.

The other possible type of imprinting effect recognized in the mouse is that based on the observation that the semi-dominant gene, fused (*Fu*), shows a differential expression according to the parental route of transmission. Both penetrance and expression are thus lower when the gene is transmitted by the mother (Reed 1937). It may be significant that *Fu* lies close to *Igf2r* on chromosome 17, because this might suggest that more than one gene in an imprinting region is subject to imprinting modification. That *Plg* and *Sod-2*, two other genes more closely linked to *Igf2r*, are not modified weakens the argument considerably, however. A further difficulty is that a lower activity of a *maternal Fu* allele is not in accord with the imprinting inactivation of the *paternal Igf2r* locus. *Fu* expression is also claimed to be subject to genetic background modification (Ruvinsky and Algulnik 1990), a characteristic that has not yet been detected with the imprinting effects seen in the genetic studies (Cattanach and Beechey 1990a).

Conclusions

The genetic studies very clearly demonstrate that it is not the whole genome, but only regions of a few chromosomes that are subject to imprinting. It is also possible that only one or a few genes within these regions are responsible for the phenomena found. It is also clear that both maternal and paternal germ lines can bring about imprinting inactivation, as best illustrated with the *Igf2* and *Igf2r* loci. The lethalities and other phenomena associated with maternal/paternal imbalance for the imprinting regions together must account for the failure of parthenogenetic and androgenetic development. Genes affecting growth during development appear to be the principal candidates for imprinting effects. However, the timing of action of these genes appears to range from preimplantation stages (possibly with proximal chromosomes 2, 6, and distal 7) to midgestation (proximal chromosome 11) and even beyond birth (distal chromosome 17).

INDICATIONS OF IMPRINTING IN HUMANS

The discovery of imprinting phenomena in mice has stimulated searches for equivalent phenomena in humans with quite impressive results (Reik 1989; Hall 1990). Some examples may ultimately prove attributable to other mechanisms, but others appear to be well established.

Hydatidiform moles

Perhaps the most dramatic example of imprinting in humans is that of the hydatidiform mole, which is regarded as a placental tumor and is usually found in pregnancies without embryonic tissue (Kajii and Ohama 1977; Lawler et al. 1982). For this reason and because these moles are usually diploid (deriving from two paternally derived haploid sets of chromosomes), the similarity to and homology with the aberrant development of androgenetic mouse embryos are evident. Similarly, ovarian teratomas have been found to have two maternal sets of chromosomes (Linder et al. 1975). Growth abnormality could be considered the key feature of these phenomena.

Triploids

Human fetal triploids show one of two distinct phenotypes, depending on whether the extra haploid set of chromosomes derives from the mother or father. Diandric triploid tissue typically comprises large cystic placentas with partial molar changes (Lawler 1984), but, if there is mosaicism, a fetus may be present and show growth retardation and some other abnormalities (Kalousek 1988). Conversely, with digynic triploidy, fetal development is severely retarded and the placentas are small, underdeveloped, and usually nonmolar. Interestingly, 3 of 15 digynic triploids studied by Jacobs et al. (1982) were molar, and all of these derived from a failure of the first meiotic division. This may be significant in terms of the imprinting process in the germ line. The similarities of the human triploids with the androgenetic and parthenogenic mouse embryos are again evident.

Uniparental disomy

Two cases of maternal disomy for chromosome 7 have been reported in humans (Spence et al. 1988; Voss et al. 1989) and may be considered comparable to disomies described in the mouse, even though isodisomy (identical copies of one chromosome) was indicated in both cases. The cases were detected because the patients suffered from cystic fibrosis (CF) inherited from their heterozygous mothers, but the key novel feature, which may be the main consequence of the disomy, was one of

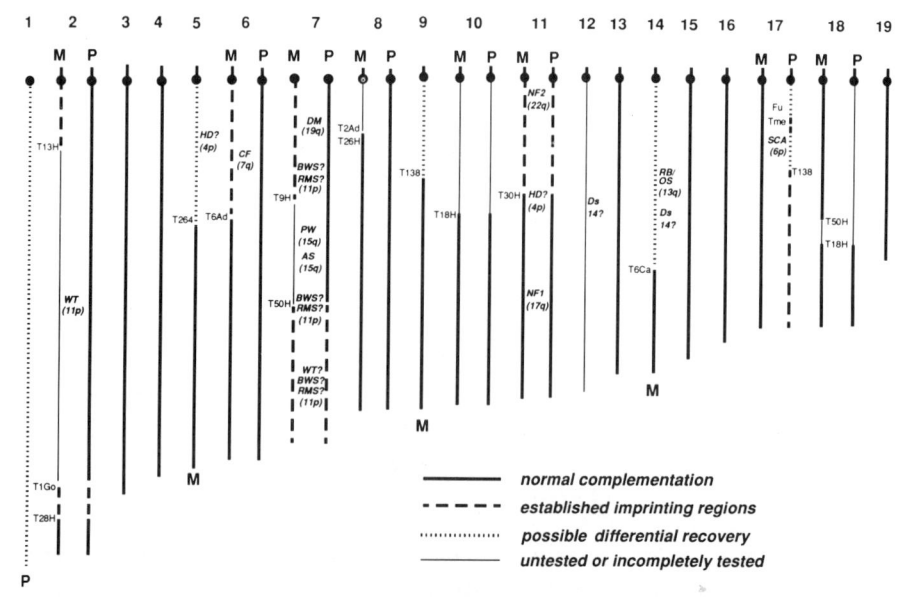

Figure 2 Imprinting map of the mouse showing probable positions of human imprinting disorders, as based on established human-mouse homologies.

short stature attributable to intrauterine growth retardation. The CF region of human chromosome 7 has homology with the proximal imprinting region of mouse chromosome 6 (Fig. 2).

A single case of human disomy for chromosome 14 has been described as showing multiple congenital malformations (Wang et al. 1990). Molecular studies indicated that both chromosomes 14 had been inherited from the father. Imprinting was suggested as the cause of the defects, but it is noteworthy that no growth effects were indicated. If genuinely due to imprinting, this case could point to homologous regions of mouse chromosome 12 or 14 as being candidates for imprinting phenomena (Fig. 2). Mouse chromosome 12 has not yet been studied; mouse chromosome 14 has shown a differential recovery effect (Cattanach and Beechey 1990b).

Deletion syndromes

The Prader-Willi (PW) and Angelman (AS) syndromes typically show "opposite" anomalous phenotypes and most commonly are associated with deletions of one region of chromosome 15 (15q11-13). The striking recent finding with PW is that the deletion almost invariably is of paternal origin, whereas with AS, the deletion derives from the mother (Knoll et al. 1989). The syndromes thus suggest that a gene within the deleted region has a differential function according to parental origin.

Consistent with this possibility are the recent findings that PW can also occur with maternal disomy (Nicholls et al. 1989) and AS can occur with paternal disomy (Malcolm et al. 1990). In addition, PW and AS have been found in a single kindred with parental origins that would be consistent with imprinting (Greenstein 1990). However, there remains cause for uncertainty; Knoll et al. (1990) have reported ten cases of non-deletion PW with no indication of uniparental disomy. It is also not certain that all PW and AS deletions occur in exactly the same region of chromosome 15.

Until recently, it was thought that the regions of chromosome 15 deleted in PW and AS had homology with a region of mouse chromosome 2, tantalizingly close to the distal imprinting region giving somewhat similar anomalous phenotypes. However, using recombinant inbred mouse strains and closely linked human probes, Nicholls et al. (1990) have established that homologs of PW and AS map to the incompletely tested central region of mouse chromosome 7 (see Figs. 1c and 2).

Expression of dominant genes

A number of inherited human disorders have been recognized as showing different levels of expression or ages of onset according to parental route of transmission. As with *Fu* in the mouse, this could be interpreted in terms of imprinting.

Huntington's disease (HD) has been suggested as a candidate gene for imprinting, as its activity appears to be dependent on parental origin (Cattanach 1986; Reik 1989). The juvenile form is most commonly inherited from the father (Ridley et al. 1988). However, a high level of inconsistency exists in this pattern, and the adult form frequently occurs among progeny of affected fathers and the juvenile form can appear among progeny of carrier mothers. This has led Laird (1990) to propose a different mechanism, based on position-effect variegation, to explain the phenomenon.

HD is located in a region of chromosome 4 (4pter–p16.2) that has two homologies (D4S43 and D4S62) with proximal mouse chromosome 5 (possibly band 5B), which may possibly show an imprinting effect (see above), but it also has homology (D4S10) with proximal mouse chromosome 11 (7B1), probably within the established imprinting region (see Figs. 1d and 2). If the latter should prove to be the true location of the HD gene in the mouse, then its pattern of inheritance and expression may yet have a basis in imprinting.

A paternal effect similar to that for HD is also evident for spinocerebellar ataxia (SCA), as the age of onset is lower when transmitted by affected fathers (Harding 1978; Zoghbi et al. 1988). SCA is located on the short arm of chromosome 6 (6p24–p21.3). This region has a number of homologies with proximal mouse chromosome 17 (17B) close to *Igf2r* at 17A2-3, which is subject to imprinting inactivation (Barlow et al. 1991), and *Fu*, which shows the parental origin effect (see Figs. 1e and 2) (Reed 1937). The direction of the effect is the same as for *Fu* (i.e., P activity > M activity) rather than as for *Igf2r* (M activity > P inactivity), suggesting that two separately and oppositely imprinting areas exist in the proximal region of mouse chromosome 17.

Maternal effects have also been found for a number of autosomal genes. The severe form (Steinert's disease) of dystrophia myotonia (DM) that is often evident at birth is usually found only in children of affected mothers (Harper 1975, 1986). On the basis of several homologies, DM at 19q13.2-q13.3 locates within the proximal imprinting region of mouse chromosome 7 at 7A (see Fig. 2). However, the direction of effect is again opposite of that expected for the region. The familial form of Beckwith-Wiedeman syndrome (BWS) also most commonly shows maternal transmission (Niikawa et al. 1986). Located at 11pter-p15.4, it maps to mouse chromosome 7 at any one of three possible positions: in the proximal imprinting region at 7B (with *Ldh*-1, *Myod*-1, and *Hras*), in the distal imprinting region of 7F (with *Th*), or more centrally at 7E (with *Calc* and *Hbb*) (see Figs. 1c and 2). Chromosome changes have been associated with sporadic cases of BWS, and two of these have shown duplications of genes locating to the mouse chromosome 7 regions *Hras*, *Igf2*, and *Hbb*. Reik (1989) has suggested that overexpression of IGF2, the human insulin-like growth factor (*Igf2* in the mouse), could cause some of the phenotypes of BWS. If this were the case, BWS would map to the distal imprinting region of mouse chromosome 7, and the direction of effect in the two species would be consistent.

Both forms of neurofibromatosis appear to show maternal effects. Neurofibromatosis von Recklinghausen (NFI) shows a greater severity of expression with maternal transmission (Miller and Hall 1978), although not invariably, and the degree of difference with maternal and paternal transmission is small. NFI at 17q11.2 would map to a distal region of mouse chromosome 11 (11D-E) outside the known imprinting region (see Figs. 1d and 2). On the other hand, neurofibromatosis type II (acoustic neuroma), which shows an earlier age of onset with maternal transmission (Eldridge 1981) is located at 22q11-q13.1. On the basis of two homologies (*Tcn-2* and *Nfh*), it might map within the proximal imprinting region of the same mouse chromosome (11A1-3) (see Figs. 1d and 2). Other human genes possibly showing imprinting effects are listed by Hall (1990).

Tumors

The genesis of tumors is complex and not well understood, but it clearly appears to involve imprinting (Reik and Surani 1989). Knudson's (1971) model of tumorigenesis has provided the framework for this concept. He proposed that tumors arise by a two-step process: The first could be gene mutation resulting in inactivation, and the second could be the specific chromosome loss associated with the tumor development that could eliminate the normal allele of the mutated gene. Inactivation of one allele by imprinting (Ponder 1988, 1990; Wilkins 1988; Reik 1989; Reik and Surani 1989) is a feasible alternative to mutation when preferential loss of a maternal or paternal chromosome is observed. This has proven to be the case for several tumors.

Nonfamilial Wilms' tumor (WT) and rhabdomyosarcoma (RMS) are among the most quoted examples for this type of scenario (Reik 1989; Scrable et al. 1989; Hall 1990; Sapienza 1990a). Individuals with small deletions of chromosome 11 may show congenital malformations other than Wilms' tumors—the so-called WAGR syndrome. The WAGR deletions at 11p13 map fairly securely to mouse chromosome 2 at bands 2E-F on the basis of four homologies (*Ly-24*, *Cas*-1, *Sey*, and *Fshb*), but, alternatively, they might possibly map to regions of mouse chromosome 7 on the basis of linkage with human *IGF2* (mouse *Igf2*). Losses of the most distal chromosome 11 markers in RMS have suggested that the gene is located at 11p15.5–pter and therefore might also map to regions of mouse chromosome 7 (see Figs. 1c and 2). Chromosome loss involving the closely linked *IGF2* locus has been suggested as a causative factor in the occurrence of these tumors (Reik 1989).

Imprinting may also be a factor in familial and sporadic retinoblastoma (RB) (Dryja et al. 1989) and sporadic osteosarcoma (OS) (Toguchida et al. 1989). The *RB* locus at 13q14.1–q14.2 maps on the basis of its homolog (*Rb*-1) to a central region of mouse chromosome 14 (14C) that may be subject to imprinting (Fig. 2).

In view of the probable role of imprinting in human tumorigenesis, it may be significant that somewhat similar findings have been found in the mouse and have involved chromosomes subject to imprinting. Bremner and Balmain (1990) have noted losses of heterozygosity attributable to structural changes in mouse chromosome 7 associated with progression of chemically induced skin tumors. Similarly, R. Cox and colleagues (pers. comm.) have found structural changes and methylation differences in chromosome 2 associated with the development of radiation-induced acute myeloid leukemias. However, the parental origins of the chromosomes showing these phenomena have not yet been ascertained.

Conclusions

The high levels of concordance between the locations of human imprinting phenomena and mouse imprinting regions are striking and strongly point to the involvement of a common mechanism. The concordance also suggests that more than one locus may be affected within the imprinting regions; only the discordance of the direction of some of the effects requires novel interpretation. Moreover, the human examples extend the range of imprinting effects beyond the major developmental effects and minor growth differentials observed in the mouse to diverse congenital abnormality. However, this may only reflect the differences in the mode of ascertainment in the two species. A greater variability of expression is also evident with human imprinting phenomena, and this may have significance for interpretation of the underlying mechanisms.

IMPRINTING AND TRANSGENES IN MICE

A number of studies have shown that the methylation patterns and degrees of expression of certain transgenes can be influenced by the parental route of transmission (Hadchouel et al. 1987; Reik et al. 1987; Sapienza et al. 1987; Swain et al. 1987; McGowan et al. 1989; DeLoia and Solter 1990). Almost invariably, these transgenes have been found to be hypomethylated and have the potential for expression in appropriate tissues when inherited from the male, but they are hypermethylated and have little, if any, expression when inherited from the female. Equally importantly, in all but one instance, this characteristic has been reversible at each generation, with the old imprint being erased and a new germ-line imprint being substituted. The size and sequence of the transgenes appear to be unimportant (Surani et al. 1988a, 1990a).

A particularly interesting case in this regard was recently described by DeLoia and Solter (1990). Integration of a transgene induced a dominantly inherited mutation causing a limb deformity. However, the transgene, and therefore the mutant phenotype, was only expressed following paternal transmission. It was concluded that the transgene had been inserted into a region of the genome that is subject to imprinting such that the endogenous gene responsible for the mutant phenotype was less transcriptionally active when transmitted from the female than when transmitted from the male. An earlier communication, however, had indicated that the transgene had been integrated into a region of chromosome 5 between *c-Kit* and *Pgm-1* (Bucan et al. 1989). This is not one of the established imprinting regions (see Fig. 2), but it is one of the regions that might show differential recoveries of maternal and paternal disomies. However, dominant genes in the vicinity have not been observed to show different expression according to parental route of transmission. The behavior of transgenes is intriguing, but there remain a number of reasons for suspecting that they may not be good models for endogenous gene imprinting.

1. Too many transgenes show parental origin effects (20%; Surani 1990a) with respect to the maximum size of the imprinting regions defined in the genetic studies (10%; Cattanach 1986; Cattanach and Beechey 1990b); this discrepancy is increasing as ongoing studies reduce the size of the regions further, possibly even to single genes. Solter (1988) and Surani et al. (1990a) suggested that the proneness of transgenes to imprinting may reflect only their mode of introduction into the genome via the male pronucleus or the design of the studies in which the methylation differences are detected.
2. None of the transgenes showing imprinting effects have yet been mapped to the imprinting regions of the genetic experiments.
3. Whereas the translocation studies and new work with *Igf2* and *Igf2r* have indicated that imprinting may result in the inactivation of either maternal or paternal copies of genes or chromosome regions, the transgenes almost invariably are inactivated or hypermethylated when maternally derived; the opposite has not yet been found.
4. Genetic background modification has not been detected either in genetic studies (Cattanach and Beechey 1990a) or in work with parthenogenetic and androgenetic embryos (Surani et al. 1990b), but the methylation and expression levels of transgenes have clearly been shown to be subject to a number of different genetic background influences, to the extent that the imprinting component can be permanently lost (Sapienza et al. 1989a,b; Reik et al. 1990; Surani et al. 1990b).

Conclusions

The numerous differences between the imprinting inactivation of transgenes and that of endogenous genes or chromosome regions warrant extreme caution in applying conclusions from one to the other. In particular, it would seem imprudent to conclude that because so many transgenes show parental origin effects, many segments of chromosomes are subject to imprinting inactivation and can inactivate neighboring sequences, or transgenes, by some form of position effect (Sapienza 1989, 1990b).

Transgene inactivation of flanking endogenous genes would seem more probable, and this is unlikely to be of any significance for gene regulation. On the other hand, the correlation between methylation level and gene expression further supports the evidence that methylation plays an important role in the regulation of genes during development (Monk and Grant 1990).

IMPRINTING OF THE X CHROMOSOME

Although most attention is currently being given to the identification of autosomal chromosome imprinting, it has been recognized for some time that the mammalian X chromosome is subject to a similar process (Lyon and Rastan 1984). Because much is known about X-inactivation, X chromosome imprinting may be expected to provide some clues for autosomal imprinting (Cattanach and Beechey 1990b).

The best-known example is the preferential inactivation of the paternal X chromosome that occurs in the extraembryonic membranes of female mice (Tagaki and Sasaki 1975; West et al. 1977; Harper et al. 1982), indicating that the maternal and/or paternal X chromosomes (X^M and X^P, respectively) receive a germ-line imprint to permit their distinction in the zygote. Unlike the autosomal imprinting phenomena so far distinguished in the genetic studies, X^P inactivation is not rigidly fixed, however. Thus, $X^P O$ mice do not inactivate their single paternal X (Frels and Chapman 1979; Papaioannou and West 1981); parthenogenetic diploid mouse embryos that survive beyond implantation may inactivate one of their two X^M chromosomes (Kaufman et al. 1978; Rastan et al. 1980), although the incidence of cells with an inactive X may be lower than normal (Endo and Tagaki 1981); and human androgenetic diploid moles with two X^P chromosomes inactivate only one (Tsukahara and Kajii 1985). However, in a recent remarkable study on mouse embryos that were trisomic for the X with one X^P and two X^M chromosomes (one rearranged), only the single X^P was found to be inactivated in extraembryonic membranes, although it could be active elsewhere (Shao and Tagaki 1990). X^P chromosomes are therefore clearly imprinted to be inactivated, whereas X^M chromosomes may be resistant.

The presence of two active X^M chromosomes appears to cause a failure of development of the extraplacental cone and extraembryonic endoderm similar to that found in parthenogenetic embryos. Significantly, Tagaki and Abe (1990) had earlier shown that the same effects on early embryonic development occurred in chromosomally unbalanced embryos containing an X^P and part of an X^M lacking the inactivation center (*Xce*) believed to be essential for X-inactivation. In these embryos, neither the X^P nor X^M was inactivated, which meant that two active X chromosomes, even though of different parental origin, could cause developmental problems. The implication of these studies was that two active X chromosomes were responsible for the earliest abnormalities of parthenogenetic embryos (Shao and Tagaki 1990). As autosomal imprinting effects (Cattanach and Beechey 1990a,b) also bring about early lethalities, the somewhat inconclusive evidence that parthenogenetic embryos with only a single X may survive marginally better than XX parthenogenotes (Mann and Lovell-Badge 1987, 1988) does not weaken the argument.

A further well-known example of nonrandom X-inactivation that may be attributed to X chromosome imprinting is the preferential X^P inactivation found in the somatic cells of marsupial females (Vandeberg et al. 1987). Less well known is the opposite minor bias toward preferential X^M inactivation that occurs in the somatic cells of female mice (Cattanach and Beechey 1990b). Other significant points pertaining to X-inactivation summarized by Cattanach and Beechey (1990b) were (1) X-inactivation is dependent on an inactivation center, (2) inactivation is achieved by the spread of an inactivating influence from the center to contiguous regions of the X, and (3) the spread of inactivation is reversible such that with age or time, reactivation of inactivated loci can occur. Moreover, the most recent evidence on X-inactivation indicates that some human X-linked loci are resistant to the X-inactivation process (Davis 1991).

Finally, imprinting pertaining to the X chromosome has been suggested as a factor in the fragile-X syndrome of humans. Laird (1987) has proposed that the disorder derives from a mutation in the X chromosome at position Xq27.3 which blocks the normal process of reactivation of the previously inactivated X that occurs in the premeiotic germ cells of females. He postulated that the persisting local inactivation is transmitted through to the next generation, where it is cytologically visible as a fragile site, and that transcriptional inhibition within the region accounts for the clinical symptoms of the disorder. Recent studies (Bell et al. 1991; Vincent et al. 1991) using probes that map close to the fragile site have demonstrated altered restriction fragment patterns in DNA digests from a proportion of fragile-X patients. These differences could result from blocking by methylation of recognition sequences. Although this tends to support Laird's (1987) hypothesis of persisting inactivation of the region, the relevance for the phenomenon of imprinting as described elsewhere in this paper remains obscure.

Conclusions

There now seems to be no good reason why X chromosome imprinting and its effects should be regarded as a different type of phenomenon from that observed with the autosomes. It is likely that the possession of two X^M chromosomes is at least in part responsible for the inviability of parthenogenotes. This might support the concept that some of the deleterious effects found with maternal or paternal autosomal disomies result from the presence of two fully active copies of vital genes, rather than two inactive or low-activity copies. As a model for autosomal imprinting, X chromosomal imprinting could suggest (1) that there is a need for only a single dose of certain autosomal genes, (2) that this requires an inactivation center that may have allelic forms, (3) that adjacent loci may be subject to a spread of inactivation, and (4) that reactivation of inactivated loci may occur with time. Indeed, the evidence from *Igf2* and *Igf2r* fully validate point 1 above; the concordance of many of the human imprinting disorders with the mouse imprinting regions clearly suggests that blocks of genes, rather than single genes, are inactivated together; and the variability of the human imprinting disorders could be understood in terms of spreading effects, allelic differences at inactivation loci, and reactivation (see points 2, 3, and 4 above).

THE ROLE OF IMPRINTING

Because certain amphibian and avian species have been found to be capable of parthenogenetic development, there has been a tendency to regard imprinting as a mammalian phenomenon. In view of the findings in *Sciara* and mealy bugs, as well as from observations in plants, this is clearly not the case. The problem appears to lie with the fact that the term imprinting is being used to describe a number of separate but connected events. The first event is the germ-line marking of segments of the genome such that maternal and paternal copies can be identified in the zygote. Strictly speaking, this is all that imprinting means. The second event is the process that causes genetic inactivation (imprinting inactivation), and this may occur by different ways in different species. Finally, there is the resulting phenotypic effect (imprinting effect) that is recognized when the parental origin of the two copies of a segment of the genome is other than the normal one maternal copy plus one paternal copy.

At this time, almost nothing is known about the nature of the germ-line imprint. It is therefore not clear whether the mechanism of imprinting is the same in the germ lines of insects, such as *Sciara*, and mammals, such as the mouse. Although it might appear from studies on X chromosomal behavior in these two species that a controlling center which can be recognized in the embryo is required, no evidence for this exists in mealy bugs.

At another level, Hultén and Hall (1990) have suggested that the differences in meiotic pairing in the germ lines of males and females may be the critical first step for imprinting in mammals. It was suggested that resultant modifications of chromatin configurations and condensation could be heritable and would form the basis for the mechanisms of subsequent differential gene expression in the embryo. However, on the basis of any such explanation, it becomes difficult to understand how there can be maternal inactivation of some loci (e.g., *Igf2*) and paternal inactivation of others (e.g., *Igf2r*).

Somewhat more is known about events associated with the genetic inactivation that takes place in the zygote, although the precise mechanisms of inactivation have not been established. Chromosome condensation, heterochromatic behavior, and asynchronous DNA replication, as well as chromosome loss, have been associated with the phenomenon, but again it is not clear whether the mechanisms are the same across all classes. Only genetic inactivation itself has so far been indicated for mammalian autosomal imprinting phenomena, but detailed chromosomal analyses of embryos at various stages of development have probably not been conducted. On the basis of both the observations with transgenes and the association between changes in methylation levels in mouse embryos and the onset of differentiation (Chapman et al. 1984; Sanford et al. 1984, 1985; Monk et al. 1987), DNA methylation remains the strongest candidate for causing imprinting inactivation. However, it is possible that methylation may not be the primary inactivation mechanism but, as with X-inactivation, may only be involved with the maintenance of the inactive state (Monk and Grant 1990). Studies on endogenous genes, which are now possible, should be informative in this regard. Whatever the precise mechanism, it is evident from *Sciara* and mouse that the original imprints must be carried through several cleavage divisions before inactivation and/or associated events (e.g., chromosome loss or heterochromatic behavior) can occur.

Although the inactivating events may take place very early in embryonic development (cf. X-inactivation), the eventual consequence defining the role of imprinting in each animal or plant group is the failure of gene expression at the appropriate time and in the appropriate tissue. In *Sciara* and mealy bugs, for example, imprinting appears to have a role in sex determination (Chandra and Brown 1975), and for the mammalian X chromosome, it has a role in dosage compensation. What then is the role of imprinting with regard to mammalian autosomes?

Hall (1990) has proposed that imprinting has evolved in mammals to restrain the growth potential of the placenta so that the female is not sacrificed to the growth of the half-foreign placenta and embryo. A more detailed model based on the same premise has been offered by Moore and Haig (1991), covering both mammals and flowering plants with the common feature of nourishing the embryo directly from maternal tissues. Their model proposes that preferential expression of a paternally derived gene in an embryo will increase the demand upon the mother favoring the embryos' survival; for the maternally derived gene, the *opposite* would be true. The model is in accord with such observations as the *overgrowth* of trophoblastic tissues in *androgenotes* and the *increased* size of mice with *paternal* disomy 11, in contrast to the opposite findings in *parthenogenotes* and *maternal* disomy 11. However, the model is not in accord with other findings such as the small size of mice with *paternal* duplication/maternal deficiency for distal chromosome

17 and the *inactivation* of the *paternal* copy of the *Igf2r* locus on proximal chromosome 17. It is also not in accord with the late postnatal onset of the growth effect observed in the studies with distal chromosome 17.

Yet there remains a broad range of evidence that growth is a common factor in most of the mouse imprinting phenomena, as well as in several of the human disorders (Cattanach and Beechey 1990a). A case can be made that the primary role of imprinting in mammals is the regulation of expression of growth factors throughout embryonic development and into early postnatal life; i.e., it is a process vital to the regulation of normal developmental processes. In addition to *Igf2* and *Igf2r*, there are numbers of growth factors, receptors, related sequences, and oncogenes in the mouse imprinting regions and homologous chromosome regions in humans that may represent candidate loci for the imprinting effects observed in the genetic studies. Some could be expressed in the placenta, but direct effects on growth of the embryo itself could also be expected.

Although growth factor regulation may be the primary function of imprinting in mammals, with X-inactivation as a model for imprinting, it might be expected that other loci located within the same segments of chromosome may also be inactivated. Should this occur by spreads of inactivation from inactivation centers (with the possibility of reactivation), then aspects of Sapienza's dominance modification concept (Sapienza 1989; Sapienza et al. 1989b) could apply. This model proposes a variability of imprinting inactivation that could help to explain the variability of some of the putative imprinting disorders in humans (e.g., Huntington's disease; see also Laird 1990). In contrast to Sapienza's model, however, most genes showing this effect would locate within the defined mouse imprinting regions, or at least on the same chromosome, rather than being scattered throughout the genome. On the basis of the evidence presently available, this would appear to be the case. Bearing in mind the recent evidence on X-inactivation, not all loci within these regions could be expected to be inactivated, however.

In conclusion, mammalian chromosome imprinting is a field that is ripe for investigation. Primary candidate genes are strongly indicated and the technology is available for determining the mechanisms by which they are regulated during embryonic development and to what extent neighboring sequences may be affected. At the basic genetic level, there is scope for identifying additional imprinting regions and further defining those that have been found. In addition, the basis for bizarre findings, such as the discordance of monozygous twins with BWS and the claimed increase in severity of MD with transmission through several generations (Hall 1990), is in need of explanation. However, the nature of the germ-line imprint that is integral to the phenomenon of imprinting as a whole may be the most difficult to resolve.

Acknowledgment

I thank Colin Beechey for his stalwart efforts in preparing the figures.

References

Anderegg, C. and C.L. Markert. 1986. Successful rescue of microsurgically produced homozygous uniparental mouse embryos via production of aggregation chimeras. *Proc. Natl. Acad. Sci.* **83:** 6509.

Barlow, D.P., R. Stoger, B.G. Herrman, K. Saito, and N. Schweifer. 1991. The mouse insulin-like growth factor type-2 receptor in imprinted and closely linked to the *Tme* locus. *Nature* **349:** 84.

Barton, S.C., M.A.H. Surani, and M.L. Norris. 1984. Role of paternal and maternal genomes in mouse development. *Nature* **311:** 374.

Barton, S.C., C.A. Adams, M.L. Norris, and M.A.H. Surani. 1985. Development of gynogenetic and parthenogenetic inner cell mass and trophectoderm tissues in reconstituted blastocysts of the mouse. *J. Embryo. Exp. Morphol.* **90:** 267.

Bell, M.V., M.C. Hirst, Y. Nakahori, R.N. MacKinnon, A. Roche, T.J. Flint, P.A. Jacobs, N. Tommerup, L. Tranebjaerg, U. Froster-Iskenius, B. Kerr, G. Turner, R.H. Lindenbaum, R. Winter, M. Pembrey, S. Thibodeau, and K.E. Davis. 1991. Physical mapping across the fragile-X: Hypermethylation and clinical expression of the fragile-X syndrome. *Cell* **64:** 861.

Berger, C.N. and C.J. Epstein. 1989. Genomic imprinting: Normal complementation of murine chromosome 16. *Genet. Res.* **54:** 227.

Bremner, R. and A. Balmain. 1990. Genetic changes in skin tumour progression: Correlation between presence of a mutant *ras* gene and loss of heterozygosity on mouse chromosome 7. *Cell* **61:** 407.

Bucan, M., J. DeLoia, J. Price, J.-L. Guenet, and D. Solter. 1989. The chromosomal localisation and molecular analysis of the transgene insertion associated with limb deformity and genomic imprinting. *Mouse Genome* **85:** 81.

Cattanach, B.M. 1986. Parental origin effects in mice. *J. Embryol. Exp. Morphol.* (suppl.) **97:** 137.

Cattanach, B.M. and C.V. Beechey. 1990a. Chromosome imprinting phenomena in mice and indications in man. *Chromosomes Today* **10:** 135.

―――. 1990b. Autosomal and X-chromosomal imprinting. In *Genomic imprinting* (ed. M. Monk and A. Surani), p. 63. The Company of Biologists, Cambridge.

Cattanach, B.M. and M. Kirk. 1985. Differential activity of maternally and paternally derived chromosome regions in mice. *Nature* **315:** 496.

Cattanach, B.M., C.V. Beechey, E.P. Evans, and M. Burtenshaw. 1991. Further localisation of the distal chromosome 2 imprinting region. *Mouse Genome* **89:** 255.

Chandra, H.S. and S.W. Brown. 1975. Chromosome imprinting and the mammalian X-chromosome. *Nature* **253:** 165.

Chapman, V., L. Forrester, J. Sanford, N. Hastie, and J. Rossant. 1984. Cell

lineage-specific under-methylation of mouse repetitive DNA. *Nature* **307**: 284.
Clarke, H.J., S. Vamuza, V.R. Prideaux, and J. Rossant. 1988. The developmental potential of parthenogenetically-derived cells in chimeric mouse embryos: Implications for actions of imprinted genes. *Development* **104**: 175.
Crouse, H.V. 1960. The controlling element in sex chromosome behaviour in *Sciara*. *Genetics* **45**: 1429.
Davis, K. 1991. The essence of inactivity. *Nature* **349**: 15.
DeChiara, T.M., A. Efstratiadis, and E.J. Robertson. 1990. A growth deficiency phenotype in heterozygous mice carrying an insulin-like growth factor II gene disrupted by targetting. *Nature* **345**: 78.
DeLoia, J.A. and D. Solter. 1990. A transgene insertional mutation of an imprinted locus in the mouse genome. In *Genomic imprinting* (ed. M. Monk and A. Surani), p.73. The Company of Biologists, Cambridge.
Dryja, T.P., S. Mukai, P. Petersen, J.M. Rapaport, D. Walton, and D.W. Yandell. 1989. Parental origin of mutations of the retinoblastoma gene. *Nature* **339**: 556.
Eldridge, R. 1981. Central neurofibromatosis with bilateral acoustic neuroma. *Adv. Neurol.* **29**: 57.
Endo, S. and N. Tagaki. 1981. A preliminary cytogenetic study of X chromosome inactivation in different parthenogenetic embryos from LT/Sv mice. *Jpn. J. Genet.* **56**: 349.
Frels, W.I., and V.M. Chapman. (1979). Paternal X-chromosome expression in extraembryonic membranes of XO mice. *J. Exp. Zool.* **210**: 553.
Fundele, R., M.L. Norris, S.C. Barton, W. Reik, and M.A. Surani. 1989. Systematic elimination of parthenogenetic cells in mouse chimeras. *Development* **106**: 20.
Fundele, R.H., M.L. Norris, S.C. Barton, M. Fehlau, S.K. Howlett, W.E. Mills, and M.A. Surani. 1990. Temporal and spatial selection against parthenogenetic cells during development of fetal chimeras. *Development* **108**: 203.
Greenstein, M.A. 1990. Prader-Willi and Angelman syndrome in one kindred with expression consistent with genetic imprinting. *Am. J. Hum. Genet.* **47**: A59.
Hadchouel, M., H. Farza, D. Simon, P. Tiollias, and C. Pourcel. 1987. Maternal inhibition of hepatitis B surface antigen gene expression in transgenic mice correlates with *de novo* methylation. *Nature* **329**: 454.
Hall, J.G. 1990. Genomic imprinting: Review and relevance to human diseases. *Am. J. Hum. Genet.* **46**: 857.
Harding, A.E. 1978. Genetic aspects of autosomal dominant late onset cerebellar ataxia. *J. Med. Genet.* **18**: 436.
Harper, M.I., M. Foster, and M. Monk. 1982. Preferential paternal inactivation in extra-embryonic tissues of early mouse embryos. *J. Embryol. Exp. Morphol.* **67**: 127.
Harper, P.S. 1975. Congenital myotonic dystrophy in Britain II. Genetic basis. *Arch. Dis. Child.* **50**: 505.
———. 1986. Myotonic disorders. In *Myology* (ed. A.G. Engel and B.Q. Bander), p. 1267. McGraw-Hill, New York.
Hughes-Schrader, S. 1948. Cytology of coccids (*Coccoidea Homoptera*). *Adv. Genetics* **2**: 127.

Hultén, M.A. and J.G. Hall. 1990. Proposed meiotic mechanisms of genomic imprinting. *Chromosomes Today* **10**: 157.

Jacobs, P.A., A.E. Szulman, J. Funkhouser, J.S. Matsuura, and C.C. Wilson. 1982. Human triploidy: Relationship between parental origin of the additional haploid complement and development of partial hydatidiform mole. *Hum. Genet.* **46**: 223.

Johnston, D.R. 1974. Hairpin-tail: A case of post-reductional gene action in the mouse egg? *Genetics* **76**: 795.

———. 1975. Further observations on the hairpin-tail (T^{hp}) mutation in the mouse. *Genet. Res.* **24**: 207.

Kajii, T. and K. Ohama. 1977. Androgenetic origin of hydatidiform mole. *Nature* **268**: 633.

Kalousek, D.K. 1988. The role of confined chromosomal mosaccism in placental function and human development. *Growth Genet. Hormones* **4**: 1.

Kaufman, M.H. 1983. *Early mammalian development: Parthenogenetic studies.* Cambridge University Press, England.

Kaufman, M.H., M. Guc-Cubrilo, and M.F. Lyon. 1978. X-chromosome inactivation in diploid parthenogenetic mouse embryos. *Nature* **271**: 547.

Kermicle, J.L. and M. Alleman. 1990. Genetic imprinting in maize in relation to the angiosperm life cycle. In: *Genomic imprinting* (ed. M. Monk and A. Surani), p. 9. The Company of Biologists, Cambridge, England.

Klar, A.J.S. 1990. Regulation of fission yeast mating-type interconversion by chromosome imprinting. In *Genomic Imprinting* (ed. M. Monk and A. Surani), p. 3. The Company of Biologists, Cambridge, England.

Knoll, J.H.M., R.D. Nichols, R.E. Magenis, J.M. Graham, M. Lalande, and S.A. Latt. 1989. Angelman and Prader-Willi syndrome share a common chromosome 15 deletion but differ in parental origin of the deletion. *Am. J. Med. Genet.* **32**: 285.

Knoll, J.H.M., K. Glatt, R.D. Nichols, S. Malcolm, and M. Lalande. 1990. Uniparental disomy not detected in Angelman syndrome. *Am. J. Med. Genet.* **47**: A225.

Knudson, A.G. 1971. Mutation and cancer: Statistical study of retinoblastoma. *Proc. Natl. Acad. Sci.* **68**: 820.

Laird, C.D. 1987. Proposed mechanism of inheritance and expression of the human fragile-X syndrome of mental retardation. *Genetics* **117**: 587.

———. 1990. Proposed genetic basis of Huntington's disease. *Trends Genet.* **6**: 242.

Lawler, S.D. 1984. Genetic studies on hydatidiform moles. *Adv. Med. Biol.* **176**: 147.

Lawler, S.D., S. Povey, R.A. Fisher, and V.J. Pickthal. 1982. Genetic studies on hydatidiform moles. II. The origin of complete moles. *Ann. Hum. Genet.* **46**: 209.

Linder, D., B. McCaw, K. Kaiser, and F. Hecht. 1975. Parthenogenetic origin of benign ovarian teratomas. *N. Engl. J. Med.* **292**: 63.

Lyon, M.F. and P.H. Glenister. 1977. Factors affecting the observed number of young resulting from adjacent-2 disjunction in mice carrying a translocation. *Genet. Res.* **29**: 83.

Lyon, M.F. and S. Rastan. 1984. Parental source of chromosome imprinting and its relevance for X-chromosome inactivation. *Differentiation* **26**: 63.

Lyon, M.F., P.H. Glenister, and S.G. Hawkes. 1972. Do the H-2 and T-loci of the

mouse have a function in the haploid phase of sperm? *Nature* **240:** 152.
Malcolm, S., M. Nichols, M., J. Clayton-Smith, S. Robb, T. Webb, J.A.L. Armour, A.J. Jeffreys, and M.G. Pembrey. 1990. Angelman syndrome can result from uniparental isodisomy. *Am. J. Hum. Genet.* **47:** A227.
Mann, J.R. and R. Lovell-Badge. 1984. Inviability of parthenogenones is determined by pronuclei, not egg cytoplasm. *Nature* **310:** 66.
———. 1987. The development of XO gynogenetic mouse embryos. *Development* **99:** 411.
———. 1988. Two maternally derived X-chromosomes contribute to parthenogenetic inviability. *Development* **103:** 129.
Mann, J.R., I. Gadi, M.L. Harbison, S.J. Abbondanzo, and C.L. Stewart. 1990. Androgenetic mouse embryonic stem cell are pluripotent and cause skeletal defects in chimeras: Implications for imprinting. *Cell* **62:** 251.
McGowan, R., R. Campbell, A. Petersen, and C. Sapienza. 1989. Cellular mosaicism in the methylation and expression of hemizygous loci in the mouse. *Genes Dev.* **3:** 1669.
McGrath, J. and D. Solter. 1984a. Completion of mouse embryogenes is requires both the maternal and paternal genomes. *Cell* **37:** 179.
———. 1984b. Inability of mouse blastomere nuclei transferred to enucleated zygotes to support development in vitro. *Science* **226:** 1317.
Metz, C.W. 1938. Chromosome behaviour, inheritance and sex determination in *Sciara*. *Am. Nat.* **72:** 485.
Miller, M. and J.A. Hall. 1978. Possible maternal effect on severity of neurofibromatosis. *Lancet* **II:** 1071.
Monk, M., and M. Grant. 1990. Preferential X-chromosome inactivation, DNA methylation and imprinting. In *Genomic imprinting* (ed. M. Monk and A. Surani), p. 55. The Company of Biologists, Cambridge, England.
Monk, M., M. Boubelik, and S. Lehnert. 1987. Temporal and regional changes in DNA methylation in the embryonic, extraembryonic and germ cell lineages during mouse embryo development. *Development* **99:** 371.
Moore, T. and D. Haig. 1991. Genomic imprinting in mammalian development: A parental tug-of-war. *Trends Genet.* **7:** 45.
Nagy, A., A. Páldi, L. Dezso, L. Varga, and A. Magyar. 1987. Parental fate of parthenogenetic cells in mouse aggregation chimeras. *Development* **101:** 67.
Nicholls, R.D., P. Neuman, and B. Horsthemke. 1990. Toward mouse models of Prader-Willi, Angelman syndromes and genetic imprinting. *Am. J. Hum. Genet.* **47:** A230.
Nicholls, R.D., J.H.M. Knoll, M.G. Butler, S. Karam, and M. Lalande. 1989. Genetic imprinting suggested by maternal heterodisomy in non-deletion Prader-Willi syndrome. *Nature* **342:** 281.
Niikawa, N., S. Ishikiriyama, S. Takahashi, A. Inagawa, H. Tohoki, Y. Ohta, and N. Hase. 1986. The Wiedeman-Beckwith syndrome: Pedigree studies on five families with evidence for autosomal dominant inheritance with variable expressivity. *Am. J. Med. Genet.* **24:** 41.
Nur, U. 1990. Heterochromatization and euchromatization of whole genomes in scale insects (*Coccoidea Homoptera*). In *Genomic imprinting* (ed. M. Monk and A. Surani), p. 29. The Company of Biologists, Cambridge, England.
Papaioannou, V.E. and J.D. West. 1981. Relationship between parental origin of the X-chromosomes, embryonic cell lineage and X-chromosomes expres-

sion in mice. *Genet. Res.* **37**: 183.
Ponder, B. 1988. Gene losses in human tumours. *Nature* **335**: 400.
―――. 1990. Inherited predisposition to cancer. *Trends. Genet.* **6**: 213.
Rastan, S., M.H. Kaufman, A.H. Handyside, and M.F. Lyon. 1980. X-chromosome inactivation in extra-embryonic membranes of diploid parthenogenetic mouse embryos demonstrated by differential staining. *Nature* **288**: 172.
Reed, S.C. 1937. The inheritance and expression of fused, a new mutation in the house mouse. *Genetics* **22**: 1.
Reik, W. 1989. Genomic imprinting and genetic disorders in man. *Trends Genet.* **5**: 331.
Reik, W. and M.A. Surani. 1989. Genomic imprinting and embryonal tumours. *Nature* **338**: 112.
Reik, W., L.K. Howlett, and M.A. Surani. 1990. Imprinting by DNA methylation: From transgenes to endogenous gene sequences. In *Genomic imprinting* (ed. M. Monk and A. Surani), p. 99. The Company of Biologists, Cambridge, England.
Reik, W., S. Collick, M.L. Norris, S.C. Barton, and M.A.H. Surani 1987. Genomic imprinting determines methylation of parental alleles in transgenic mice. *Nature* **328**: 248.
Ridley, R.M., C.D. Firth, T.J. Crow, and P.M. Conneally. 1988. Anticipation in Huntingtons disease is inherited through the male line but may originate in the female. *J. Med. Genet.* **25**: 589.
Ruvinsky, A.O. and A.I. Agulnik. 1990. Genetic imprinting and the manifestation of the fused gene in the house mouse. *Dev. Genet.* **11**: 263.
Sanford, J.P., V.M. Chapman, and J. Rossant. 1985. Selective silencing of eukaryotic DNA. *Science* **189**: 426.
Sanford, J.P., L. Forrester, and V. Chapman. 1984. Methylation patterns of repetitive DNA sequences in germ cells of *Mus* musculus. *Nucleic Acids Res.* **12**: 2823.
Sapienza, C. 1989. Genome imprinting and dominance modification. *Ann. N.Y. Acad. Sci.* **564**: 24.
―――. 1990a. Parental imprinting of genes. *Sci. Am.* **263**: 52.
―――. 1990b. Sex linked dosage-sensitive modifiers as imprinting genes. In *Genomic imprinting* (ed. M. Monk and A. Surani), p. 107. The Company of Biologists, Cambridge, England.
Sapienza, C., J. Paquette, T.N. Tran, and A. Peterson. 1989a. Epigenetic and genetic factors affect transgene methylation imprinting. *Development* **107**: 165.
Sapienza, C.A., A.C. Peterson, J. Rossant, and R. Balling. 1987. Degree of methylation of transgenes is dependent on gamete of origin. *Nature* **328**: 251.
Sapienza, C., T.H. Tran, J. Paquette, R. McGowan, and A. Peterson. 1989b. A methylation mosaic model for mammalian genome imprinting. *Prog. Nucleic Acid Res. Mol. Biol.* **36**: 145.
Scrable, H., W. Cavenee, F. Ghavimi, M. Lovell, K. Morgan, and C. Sapienza. 1989. A model for embryonal rhabdomyosarcoma tumorigenesis that involves genome imprinting. *Proc. Natl. Acad. Sci.* **86**: 7480.
Searle, A.G. and C.V. Beechey. 1978. Complementation studies with mouse translocations. *Cytogenet. Cell. Genet.* **20**: 282.

———. 1985. Noncomplementation phenomena and their bearing on nondisjunctional events. In *Aneuploidy* (ed. V.L. Dellarco et al.), p. 363. Plenum Press, New York.
———. 1990. Genome imprinting phenomena on mouse chromosome 7. *Genet. Res.* **56:** 237.
Searle, A.G., C.E. Ford, and C.V. Beechey. 1971. Meiotic non-disjunction in mouse translocations and the determination of centromere position. *Genet. Res.* **18:** 215.
Senior, P.V., S. Byrne, W.A. Brammar, and F. Beck. 1990. Expression of the IGF-II/mannose-6-phosphate receptor mRNA and protein in the developing rat. *Development* **109:** 67.
Shao, C. and Tagaki, M. 1990. An extra maternally derived X-chromosome is deleterious to early mouse development. *Development* **110:** 969.
Snell, G.D. 1946. An analysis of translocations in the mouse. *Genetics* **31:** 151.
Spence, J.E., R.G. Perciaccante, G.M. Greig, H.F. Willard, D.H. Ledbetter, J.F. Hejtmancik, M.S. Pollack, W.E. O'Brien, and A.L. Beaudet. 1988. Uniparental disomy as a mechanism for human genetic disease. *Am. J. Hum. Genet.* **42:** 217.
Solter, D. 1988. Differential imprinting and expression of maternal and paternal genomes. *Annu. Rev. Genet.* **22:** 127.
Surani, M.A.H. and S.C. Barton. 1983. Development of gynogentic eggs in the mouse: Implications for parthenogentic embryos. *Science* **222:** 1034.
Surani, M.A.H., S.C. Barton, and M.L. Norris. 1984. Development of reconstituted mouse eggs suggest imprinting of the genome during genetogenesis. *Nature* **308:** 548.
———. 1986a. Nuclear transplantation in the mouse: Heritable differences between parental genomes after activation of the embryonic genome. *Cell* **45:** 127.
———. 1987. Influence of parental chromosomes on spatial specificity in androgenetic-parthenogenetic chimaeres in the mouse. *Nature* **326:** 395.
Surani, M.A.H., W. Reik, and N.D. Allen. 1988a. Transgenes as molecular probes for genomic imprinting. *Trends Genet.* **4:** 59.
Surani, M.A.H., S.C. Barton, S.K. Nowlett, and M.L. Norris. 1988b. Influence of chromosomal determinants on development of androgenetic and parthenogenetic cells. *Development* **103:** 171.
Surani, M.A.H., W. Reik, M.L. Norris, and S.C. Barton. 1986. Influence of germ-line modifications of homologous chromosomes on mouse development. *J. Embryol. Exp. Morphol.* (suppl.) **97:** 123.
Surani, M.A., N.D. Allen, S.C. Barton, R. Fundele, S.K. Howlett, M.L. Norris, and W. Reik 1990a. Developmental consequences of imprinting of parental chromosomes by DNA methylation. *Philos. Trans. R. Soc. Lond. B* **326:** 313.
Surani, M.A., R. Kothary, N.D. Allen, P.B. Singh, R. Fundele, A.C. Fergusen-Smith, and S.C. Barton. 1990b. Genomic imprinting and development in the mouse. In *Genomic imprinting* (ed. M. Monk and A. Surani), p. 89. The Company of Biologists, Cambridge. England.
Swain, J.L., T.A. Stuart, and P. Leder. 1987. Parental legacy determines methylation and expression of an autosomal transgene: A molecular mechanism for parental imprinting. *Cell* **50:** 719.
Tagaki, N. and K. Abe. 1990. Detrimental effects of two active X chromosomes

on early mouse development. *Development* **109**: 189.
Tagaki, N. and M. Sasaki. 1975. Preferential expression of the paternally derived X chromosome in the extraembryonic membranes of the mouse. *Nature* **256**: 640.
Tease, C. and B.M. Cattanach. 1986. Mammalian cytogenetic and genetic tests for autosomal non-disjunction and chromosome loss in mice. In *Chemical mutagens: Principles and methods for their detection* (ed. F.J. de Serres), vol. 10, p. 218. Plenum Press, New York.
Toguchida, J., K. Ishizaki, M.S. Sasaki, Y. Nakamura, M. Ikenaga, M. Kato, M. Sugimot, Y. Kotoura, and T. Yamamuro. 1989. Preferential mutation of paternally derived RB gene as to initial event in sporadic osteosarcoma. *Nature* **338**: 156.
Tsukahara, M. and T. Kajii. 1985. Replication of X-chromosomes in complete moles. *Hum. Genet.* **71**: 7.
Vandeberg, T.L., E.S. Robinson, P.S. Samollow, and P.G. Johnson. 1987. X-linked gene expression and X-chromosome inactivation: Marsupials, mouse and man compared. *Isozymes Curr. Top. Biol. Med. Res.* **15**: 225.
Vincent, A., D. Heitz, C. Petit, C. Kretz, I. Oberle, and J.-L. Mandel. 1991. Abnormal pattern detected in fragile-X patients by pulsed-field gel electrophoresis. *Nature* **349**: 624.
Voss, R., E. Ben-Simon, A. Avital, Y. Zlotogora, J. Dagan, S. Godfry, Y. Tikochinski, and J. Hillel. 1989. Isodisomy of chromosome 7 in a patient with cystic fibrosis: Could uniparental disomy be common in humans? *Am. J. Hum. Genet.* **45**: 373.
Wang, J.-C., M.B. Passage, P. Yen, L.J. Shapiro, and T.K. Mohandas. 1990. Uniparental heterodisomy for chromosome 14 in a phenotypically abnormal familial 13/14 Robertsonian translocation carrier. *Am. J. Hum. Genet.* **47**: A99.
West, J.D., W.I. Freis, V.M. Chapman, and V.E. Papaioannou. 1977. Preferential expression of the maternally derived X-chromosome in the mouse yolk sac. *Cell* **12**: 873.
Wilkins, R.J. 1988. Genomic imprinting and carcinogenesis. *Lancet* I: 329.
Winking, H. and L.M. Silver. 1984. Characterization of a recombinant mouse t-haplotype that expresses a dominant lethal maternal effect. *Genetics* **108**: 1013.
Zoghbi, H.Y., M.S. Pollack, L.A. Lyons, R.E. Ferrell, S.P. Diagner, and A.L. Beaudet. 1988. Spinocerebellare ataxia; variable age of onset and linkage to HLA in a large kindred. *Ann. Neurol.* **23**: 580.

The X-inactivation Center and Mapping of the Mouse X Chromosome

Stephen D.M. Brown

Department of Biochemistry and Molecular Genetics
St. Mary's Hospital Medical School
London W2 1PG England

The discovery, nearly 30 years ago, of the phenomenon of X chromosome inactivation, whereby one of the two X chromosomes in female mammals is inactivated, was a milestone in mammalian genetics. Nevertheless, the genetic pathways underlying the developmental switch of X-inactivation are still not fully resolved. Several lines of evidence have indicated the presence of a controlling center on both the mouse and human X chromosomes, from which X chromosome inactivation is initiated. The recent growth of technologies applicable to the long-range mapping of mammalian chromosomes has enabled the mapping and cloning of sequences from those chromosomal regions identified as containing the X-inactivation center. The long-range mapping of the X-inactivation center region will allow us to test the relevant underlying sequence regions for functional X-inactivation activity and to begin to assess the role and mechanism of the X-inactivation center.

This chapter describes:

❑ the role of initiation in X chromosome inactivation

❑ the chromosomal location of the X-inactivation center in mouse and human species

❑ techniques employed in the molecular and genetic mapping of the mouse X chromosome

❏ the long-range genetic and physical mapping of the mouse X-inactivation center region

❏ the biological assay of sequences from the X-inactivation region for functional inactivation activity

INTRODUCTION

Since the phenomenon of X-inactivation in female mammals was first proposed (Lyon 1961), the genetic basis for the switch controlling the decision to inactivate one or the other of the two X chromosomes has remained an enigma. In female mammals, X-inactivation of the two X chromosomes occurs at random in the embryonic lineages (Lyon 1988). The switch is developmentally controlled and occurs at 5.5 days postcoitum in the mouse embryo proper, whereas in the extraembryonic lineages, X-inactivation occurs at an earlier stage (Kratzer and Gartler 1978; Monk and Harper 1979). This apparent correlation between differentiation and the determination of a switch to X-inactivation is further supported by the knowledge that many embryonal carcinoma (EC) and embryonic stem (ES) cell lines possess two active X chromosomes; X-inactivation proceeds as EC or ES cell lines are induced to differentiate (Tagaki and Martin 1984; Rastan and Robertson 1985). A further developmental switch in the process of controlling X-inactivation occurs during oogenesis, where reactivation of the formerly inactivated X chromosome must occur; the inactive X chromosome is reactivated at about the time of onset of meiosis (Kratzler and Chapman 1981).

X-inactivation arrests the genetic activity of most of the genes on the inactivated chromosome—a state that is stably inherited within any cell lineage. The inactivation of genes appears to be controlled at the level of transcription (Graves and Gartler 1986; Nadon et al. 1988; Brown et al. 1990) and is associated with late replication (Schmidt and Migeon 1990). Significantly, some genes on both the short and long arms of the human X chromosome have been shown to escape X-inactivation, indicating that the process is not a global phenomena but decidedly sequence-specific (Shapiro et al. 1979; Goodfellow et al. 1984; Brown and Willard 1989a; Sckneider-Gadicke et al. 1989; Fisher et al. 1991).

Mechanisms of X-inactivation

In exploring the possible mechanisms of control of X-inactivation, we must consider the kinds of processes and switches that might lead to the state of X-inactivation. Three phases can be considered for the entire process. First, there is a phase where a decision or switch pathway operates to initiate the process of X-inactivation and to identify which of the two X chromosomes will be inactivated. Second, this information must be transferred or "spread" in *cis* to most genes on that chromosome; in addition, this information transfer must imprint the chromosome in a way such that X-inactivation is maintained within a cell lineage. Finally, the process must be reversible so that the inactive X chromosome can be reprogrammed to an active state during the production of oocytes. All of these phases may be initiated by a specific locus or center on the X chromosome, but that of initiation and spreading is the most approachable by genome mapping.

Initiation and spreading

Very early experiments involving X-autosome translocations in mice (Russell 1963; Russell and Montgomery 1970) or autosomal insertions into the X chromosome (Cattanach 1966) indicated that autosomal genes for coat color placed adjacent to X chromosome material could be inactivated, resulting in coat color variegation. In the case of X-autosome translocations, only autosomal genes involved in one of the reciprocal translocation products were affected. These observations suggested (1) the existence of a center (or centers) on the X chromosome from which X-inactivation could be initiated and (2) a mechanism of spreading by which the effect could be transferred into adjacent autosomal regions. It is unclear whether the inactivation of adjacent autosomal regions reflects autosomal genes coming under mechanisms of inactivation similar to those of X-linked genes or whether it reflects a general position effect resulting from the presence of neighboring inactive heterochromatin. Indeed, subsequent reactivation of the autosomal genes in the X-autosome insertion Is(In7;X)1Ct (Cattanach 1974) suggests that the autosomal loci are under less stringent maintenance than are the X-linked genes. Nevertheless, Lyon et al. (1986) have shown that there is some information transfer or spread in *cis* along the X chromosome from a center that determines the inactivated state for X-linked genes. Using two X-autosome translocations, T37H and T38H, that lie distal and proximal, respectively, to the ornithine transcarbamylase (*Oct*) locus, these authors were able to demonstrate genetically that when *Oct* is separated from the rest of the X chromosome, it fails to undergo X-inactivation.

Focus on the supposed phenomenon of spreading appears to have arisen from the observations of inactivation of autosomal material adjacent to X-linked material, where the extent of inactivation of autosomal material seemed to vary and to be related to the distance from the translocation breakpoint (Russell 1963; Eicher 1970; Russell and Montgomery 1970; Cattanach 1974). However, as indicated above, it is unclear whether this is related to the properties of inactivated genes peculiar to autosomal segments or reflects a general property of genes (including X-linked genes) that are subject to X-inactivation. The mechanism of information transfer from an inactivation center to X-linked genes may be simultaneous and instantaneous and distinct from a mechanism that involves spreading at some rate along the chromosome. Assessments of the mechanism of information transfer on the human X chromosome indicate that spreading may not be in effect. As is discussed in detail below, a human X-inactivation center appears to be located on the proximal long arm at Xq13; human X chromosomes deleted in this region fail to inactivate. If a mechanism of spreading is in operation, loci on the short arm are inactivated by spreading of the X-inactivation across the centromere into the short arm. The operation of a spreading mechanism would appear to be less efficient on the short arm as indicated by the presence of a number of genes that escape X-inactivation, including *XG*, *STS*, *MIC2*, and *ZFX*. *TIMP*, which is located in band Xp11, is inactivated (Brown et al. 1990) and may represent a boundary for the inactivated region on the short arm.

The idea of an inactivation boundary can be disputed, however, by two recent observations demonstrating that genes escaping inactivation are interspersed among genes that are inactivated. The first evidence is that a gene encoding ribosomal protein S4, mapping at Xq13 close to the inactivation center on the long arm of the human X chromosome, escapes X-inactivation (Fisher et al. 1990). The second observation relates to the analysis of gene expression on a duplication-deficient X chromosome, rec(X). The rec(X) chromosome, Xqter-Xq26.3::Xp22.3-Xqter, has a deletion of Xp22.3 to Xpter and a duplication of the Xqter to Xq26.3 region and retains some of the loci (e.g., *STS* and *MIC2*) on the distal short arm, Xp, that fail to undergo inactivation (Mohandas et al. 1987). In addition, it carries a duplication of the segment Xqter to Xq26, including the normally inactivated locus *G6PD*, attached to the tip of the short arm. Analysis of inactivated rec(X) chromosomes indicates that *STS* and *MIC2* remain active but that the duplicated *G6PD* gene lying distal to the active region in Xp is inactivated. Thus, inactivation can spread through a region containing genes not subject to X-in-activation.

These results emphasize the sequence-specific nature of the X-inactivation process and indicate that any mechanism of information transfer to X-linked genes does not involve a mechanism of facilitated spread that requires inactivation of intervening genes or is distance-dependent. Rather, the mechanism could equally be direct and involve the juxtaposition of regulatory sequences at the promoter of a gene with the inactivation center; only genes carrying the requisite regulatory sequences would be subject to X-inactivation (McBurney 1988). However, it should be pointed out that such control elements common to inactivated X-linked genes have not been discovered. Nevertheless, these ideas underscore the possible pivotal role of an X-inactivation center and the necessity to map its detailed location on the X chromosome and determine its structure and the nature of its interaction with X-linked genes.

CHROMOSOMAL LOCATION OF THE X-INACTIVATION CENTER

Mouse inactivation center

The evidence for the chromosomal location of the mouse X-inactivation center comes from two analyses of X-autosome translocations and X chromosome deletions. As considered above, translocations of portions of the X chromosome to autosomes indicate the presence of an X-inactivation center. Cytogenetic evidence based on either late replication or Kanda staining indicated that only one of the reciprocal translocation products could undergo X-inactivation (Russell and Cacheiro 1978; Rastan 1983). For example, in an unbalanced embryo X^n (normal X), X^{16}, 16, 16 carrying a translocation at the T16H breakpoint, inactivation of the X^{16} product was not observed, even though it would have restored genetic balance to the embryo. Equally, in an X^n, X^n, 16^X, 16 unbalanced embryo also carrying a translocation at the T16H breakpoint, which would be expected to carry three X-inactivation centers, two chromosomes were inactivated in each cell. As would be predicted, the 16^X product was inactivated, indicating the presence of a single X-inactivation center that lies distal to the T(X;16)16H breakpoint (Rastan 1983).

The cytogenetic location of the mouse X-inactivation center is presented in Figure 1. From analyses of the inactivation of other X-autosome translocation products, the X-inactivation center appears to lie between the T(X;16)16H breakpoint and the T(X;7)6R1 breakpoint (Fig. 1). The position of the mouse X-inactivation center was further refined by analysis of X chromosome deletion derivatives of female ES cell lines (Rastan and Robertson 1985). A number of ES cell lines were derived in which one of the two X chromosomes was deleted for a distal region of varying sizes. Each cell line appeared to be stable, and the cytogenetic breakpoints based on Giemsa (G)-banded karyotyping were determined (Robertson et al. 1983). Following differentiation, six lines were analyzed by Kanda staining for the presence of inactivated chromosomes. The HD3 cell line demonstrated a large fraction of metaphases with dark-staining heterochromatic chromosomes, as did two other cell lines with breakpoints distal to HD3. Cell lines with breakpoints proximal to HD3 showed no dark-staining chromosomes, indicat-

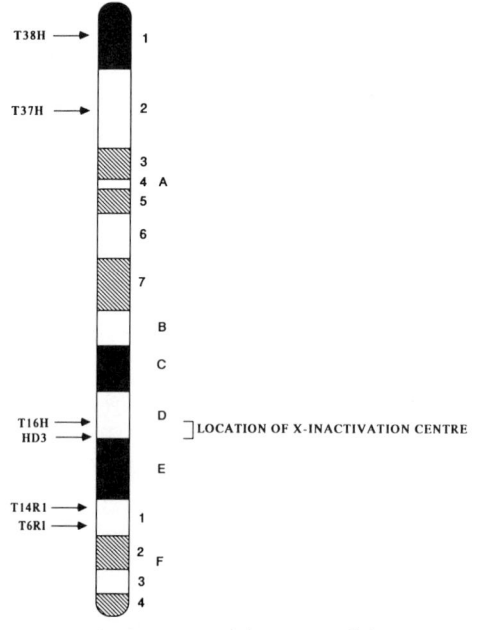

Figure 1 Cytogenetic location of the mouse X-inactivation center. Breakpoints of a variety of mouse X-autosome translocations important in the analysis of the location of the mouse X-inactivation center are indicated on the physical G-banded map of the mouse X chromosome.

ing that they possessed only one X-inactivation center on the normal X chromosome and consequently no inactivated chromosomes. It follows that the X-inactivation center lies proximal to the HD3 breakpoint on the mouse X chromosome.

The HD3 breakpoint has been localized cytogenetically by G-banding to the border of band D/E (Fig. 1). The T(X;16)16H breakpoint has been localized cytogenetically to the distal quarter of band D. Thus, the mouse X-inactivation center appears to lie in the distal portion of band D on the X chromosome and in the vicinity of the Ta locus that lies just distal to the T16H breakpoint. A cytogenetically undetectable deletion, Ta^{25H} (Cattanach et al. 1989b), circumscribes the location of the X-inactivation center. Male mice carrying Ta^{25H} exhibit the testicular feminization (Tfm) phenotype and the tabby (Ta) phenotype. The mutation appears to be a deletion which involves the closely linked Tfm and Ta loci (Cattanach et al. 1991). Nevertheless, female mice heterozygous for the Ta^{25H} mutant show the expected mosaic phenotype, indicating that X-inactivation is unaffected on the Ta^{25H} chromosome, thereby excluding the Ta^{25H} deletion from carrying the X-inactivation center.

In addition to the cytogenetic data pertaining to the location of the mouse X-inactivation center, genetic data are also available. Cattanach has investigated the *Xce* locus on the mouse X chromosome that appears to influence the probability of X chromosome inactivation (Johnston and Cattanach 1981). Three alleles have been observed, and each allele appears to act by influencing the likelihood that the X chromosome carrying the allele is inactivated. In Xce^a/Xce^b heterozygotes, the chromosome carrying Xce^a is preferentially inactivated; in Xce^b/Xce^c heterozygotes, the Xce^b chromosome is preferentially inactivated. Mapping of the *Xce* alleles (Cattanach et al. 1989a) has indicated that the locus appears to be inseparable by recombination from *Ta* and is therefore close to the mouse X-inactivation center. Whether *Xce* alleles represent alleles of the X-inactivation center per se or a second locus that influences the action of the X-inactivation center is unclear. If the latter is the case, then the genetic base for the initiation of X-inactivation may be more complex than previously thought and may involve the interaction of at least two genetic loci.

Human X-inactivation center

Studies of a variety of X chromosome deletions and X-autosome translocations in humans has allowed the delineation of a putative X-inactivation center at Xq13 on the long arm of the human X chromosome (Allderdice et al. 1978; Mattei et al. 1981; Tabor et al. 1983). The center appears to lie proximal to a large interstitial deletion of the long arm whose proximal breakpoint is in Xq13 (Tabor et al. 1983). In addition, the center appears to lie distal to the breakpoint of an (X;14) translocation in Xq13 (Allderdice et al. 1978). Molecular analysis of the (X;14) translocation products indicates that whereas the active X^{14} translocation product retains the *AR* (Lubahn et al. 1988) and *CCG1* loci (Brown et al. 1989), the inactivation center and the *PGK1* locus have been translocated to the 14^X product (Brown and Willard 1989b). This would indicate that the human X-inactivation center lies distal to the *AR* and *CCG1* loci on the human X chromosome and in the vicinity of *PGK1*.

Further analysis of a number of hybrids carrying translocation breakpoints from this region has defined a small interval, distal to the *RPS4X* and *PHKA* loci, that appears to define the human X-inactivation center region (Brown et al. 1991b). It is to this region that the *XIST* human cDNA, which is expressed only on inactive X chromosomes, has recently been mapped (Brown et al. 1991a,b). A murine homolog, *Xist*, also appears to be expressed only on the inactive X chromosome in mouse (N. Brockdorff et al., unpubl.). In addition, its genetic localization in the mouse, just distal of *Phka*, is conserved with respect to the human X chromosome (see Fig. 4) (R. Hamvas et al., unpubl).

Comparative location of mouse and human X-inactivation centers

It is clear from the accompanying analysis that the mouse and human X-inactivation centers appear to map to a roughly equivalent location within a conserved linkage group on their respective X chromosomes (Fig. 2). The human X-inactivation center lies at the proximal end of a large linkage group that includes *AR*, *PGK1*, and *PLP*, whose order is conserved on the mouse X chromosome. On the mouse X chromosome, the proximal limit of this conserved linkage group lies somewhere between the closely linked *Zfx* and *Ar* (*Tfm*) loci (Mitchell et al. 1989), as *Zfx* maps to the short arm of the human X chromosome. The mouse X-inactivation center would be expected to lie in the region surrounding the *Ar* and *Ta* loci, distal to the T16H breakpoint.

MOLECULAR MAPPING OF THE MOUSE X CHROMOSOME

To isolate and characterize sequences involved in X-inactivation on the mouse X chromosome, it has been necessary to lay the framework for a detailed molecular map of the mouse X chromosome that offers access to all of the sequences in the relevant chromosomal region. Ultimately, as is discussed below, identification of the X-inactivation center will depend on a biological assay for inactivating sequences. To achieve this, all of the relevant sequences must be available for testing. It was therefore necessary to create a detailed genetic map of molecular markers in the central region of the mouse X chromosome in order to form the basis for construction of long-range physical maps and eventually for the establishment of contigs of clones covering the entire region.

Recovering mouse X chromosome clones: Microdissection and microcloning

The most direct method for recovering DNA clones from individual chromosomes involves the physical microdissection of the relevant chromosome region (Brown 1985, 1991; Brown et al. 1988), followed by microcloning of the collected DNA. This technique involves dissecting the appropriate chromosome region using fine micromanipulative procedures and collecting a number of chromosome fragments. Subsequently, purified DNA from the collected fragments is digested and ligated into an appropriate cloning vector using microprocedures. Classically, the technique involved dissection from unstained chromosomes observed under phase-contrast microscopy, but, recently, it has been improved to incorporate finer dissections from banded chromosome ma-

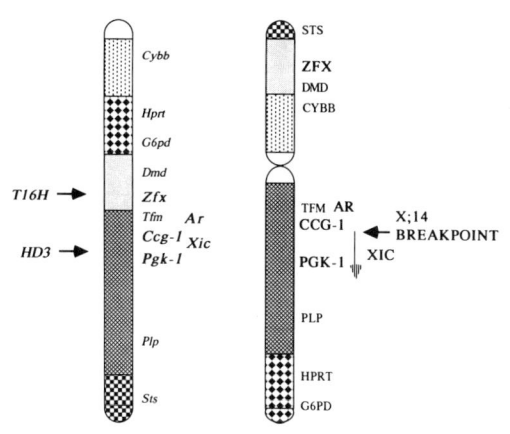

Figure 2 Comparative maps of the mouse and human X chromosomes. Relative positions of the five major linkage groups conserved between the human and mouse X chromosomes are indicated by different shadings. Loci important in determining the relative location of the mouse (*left*) X-inactivation center (*Xic*) and the human (*right*) X-inactivation center (*XIC*) are indicated in boldface type.

terial and the use of polymerase chain reaction (PCR) amplification to improve clone recovery (Ludecke et al. 1989).

Microdissection and microcloning were successfully used to obtain a large number of clones from the central span of the mouse X chromosome (Fisher et al. 1985) and from the whole mouse X chromosome (Brockdorff et al. 1987a). Dissection was from unstained metaphase chromosomes observed under phase contrast, performed on a wild mouse (CD) karyotype in which all chromosomes but the X, 19, and Y were fused as large metacentrics, leaving the mouse X clearly visible as the only large acrocentric chromosome. For dissection over the central span of the mouse X chromosome, 550 microclones were recovered (Fisher et al. 1985). A dissection of the whole mouse X chromosome gave 2000 microclones (Brockdorff et al. 1987a), most of which were of small insert size (≤400 bp). The unusual distribution of the recovered genomic fragments is caused by acid fixation of metaphase chromosomes prior to microdissection and the preferential acid hydrolysis of larger restriction fragments that prevents their cloning (Brown and Greenfield 1987). Acid hydrolysis has been a major impediment to the PCR amplification of microdissected material, and unusual steps, including the use of single-cell metaphase preparative techniques, had to be introduced to avoid lengthy acid treatment and to be able to amplify the microdissected chromosome material (Ludecke et al. 1989). Nevertheless, classical microdissection and microcloning produced usable libraries of X chromosome microclones that, in the case of the regional dissection, were prelocalized to the central region of the mouse X chromosome. The vast majority of clones recovered from these microclone banks were X-chromosome-specific (Fisher et al. 1985; Brockdorff et al. 1987a).

Recovering mouse X chromosome clones: Flow-sorting and linking libraries

In addition to microdissection and microcloning, two other techniques have been critical in preparing clone banks of the mouse X chromosome. The first technique involves flow sorting and the preparation of libraries constructed from DNA recovered from fluorescently stained chromosomes that have been separated on a flow cytometer according to DNA content and base composition. In the case of the X chromosome, flow sorting of chromosomes from the same CD karyotype that was used for microdissection of the X chromosome produced 10^6 clones with an average insert size of 3 kb (Amar et al. 1985). However, only 25-30% of the recovered clones were derived from the X chromosome. Disteche et al. (1982) also used flow sorting of chromosomes from mice carrying Cattanach's translocation to generate X chromosome clones, although this approach generates a mixed population of chromosome 7 and X chromosome clones in the clone bank.

A second approach that has been particularly successful in generating new probes to the mouse X chromosome has been the production of linking libraries (Brockdorff et al. 1990). Linking clones (Poutska and Lehrach 1986) span rare-cutter restriction sites that are a feature of the methylation-free CpG islands, 5' to most transcribed sequences. CpG islands, 1-2 kb in length, occur approximately every 100 kb in the genome, and a variety of restriction enzymes that recognize unmethylated, CpG-rich sequences cut preferentially within the island sequences and thus infrequently within the genome (Lindsay and Bird 1987). Linking clones, spanning CpG islands as they do, are particularly useful for long-range physical mapping (see below) but also have the advantage that they are likely to include the 5' region of coding sequences.

Linking clones are derived by recircularization of large partial digest fragments of total genomic DNA, followed by digestion with an appropriate rare-cutter restriction enzyme (Smith et al. 1987). Only molecules that contain the appropriate restriction site are linearized and cloned. For the mouse X chromosome, linking clones for two rare-cutter restriction enzymes, *Not*I and *Eag*I, were derived. *Not*I is one of the most infrequent cutters, recognizing sequences in only 10% of the CpG islands (Lindsay and Bird 1987). *Eag*I cuts ten times as frequently, with most CpG islands appearing to contain a site. For the mouse X chromosome representing 6% of the mouse genome, 200 *Not*I sites and 2000 *Eag*I sites could be expected to be present. To prepare representative linking libraries to the mouse X chromosome, DNA from a human-mouse somatic cell hybrid, C1.8, containing the X as the only mouse chromosome, was partially digested and recircularized. The recircularized molecules were linearized with either *Not*I or *Eag*I, and linearized molecules were cloned into the bacteriophage λ EMBL3A*Not* vector. Recovered clones were screened with total mouse DNA to identify mouse sequences. Overall, 70 *Not*I clones and more than 200 *Eag*I clones were recovered (Brockdorff et al. 1990). Many of these clones were confirmed as being mouse X-chromosome-specific and were initially localized to the mouse X chromosome using a variety of somatic cell hybrids containing differing X-autosome translocations (Avner et al. 1987b). In particular, a large number of linking clones were localized in the region of the X-inactivation center between the T16H and T14R1 breakpoints.

Interspecific mouse backcrosses

To map genetically the available DNA markers to the mouse X chromosome, interspecific mouse backcrosses (Brown 1985; Roberts et al. 1985) have become the method of choice. A cross between female laboratory mice (*Mus domesticus*) and male wild mice (*Mus spretus*) gives fertile F_1 females that can be backcrossed to laboratory mice. The advantage of such a cross is the high evolutionary divergence of the parental species and the resultant ease of detection of restriction-fragment-length variants (RFLVs) for even the shortest DNA probes. Of 39 X chromosome microclones examined with the restriction enzymes *Taq*I and *Msp*I, only 3 failed to demonstrate RFLVs between *M. domesticus* and *M. spretus* (Brockdorff et al. 1987b). For this reason, the interspecific backcross is multipoint in nature—each backcross progeny can be scored for the RFLVs of the majority of clones. Examination of the segregation of RFLVs for each clone in the backcross progeny allows us not only to determine the genetic distance between pairs of DNA markers, but also to ascertain the genetic order of markers across the chromosome. Genetic analysis of interspecific backcrosses to date has indicated that over short distances (<25 cM), double recombinants are relatively rare and triple recombinants are rarer still (Brockdorff et al. 1987a; Ceci et al. 1989; Seldin et al. 1989). Thus, gene order between available markers in individual chromosome regions is readily determined by minimizing the observed number of recombinants. Furthermore, the analysis of a large number of DNA markers over a chromosome region delineates a number of small genetic intervals defined by adjacent crossover events. A simple pedigree analysis in a limited panel of mice carrying the appropriate crossovers allows the rapid assignment of a new clone to a particular genetic interval (Avner et al. 1987a; Brockdorff et al. 1987a,b, 1988). In this way, groups of tightly linked clones can readily be circumscribed, providing valuable information for subsequent linkage in a physical map (see below).

A number of interspecific mouse backcrosses have been analyzed for X chromosome markers (Amar et al. 1985, 1988; Avner et al. 1987a; Brockdorff et al. 1987a,b, 1988; Cavanna et al. 1988; Mullins et al. 1988, 1990; Ryder-Cooke et al. 1988; Disteche et al. 1989; Herman and Walton 1990). Most of these crosses also carried visible mutations or protein polymorphisms for well-characterized loci (Table 1) that act as anchor points and aid in the conjunction of the molecular genetic map with the classical genetic map of the mouse X chromosome (Lyon and Searle 1989). In addition, the exchange of a number of markers between laboratories involved in the mapping of the mouse X chromosome has allowed considerable alignment of the relative locations of probes that were exclusive to the maps originating from a single laboratory. One map, determined from the analysis of two interspecific crosses segregating for the *Hq* and *Ta* loci (Brockdorff et al. 1987a,b) and the *mdx* locus (Cavanna et al. 1988), establishes the order of 40 molecular markers across the mouse X chromosome (Fig. 3). Recently, the X chromosome committee of the International Mouse Genome Mapping Workshop produced a composite map of molecular markers across the mouse X chromosome. This map includes the most likely order of molecular markers, their genetic distance apart, and, where known, their physical linkage and physical distance apart. Overall, approximately 100 DNA probes have been mapped to the mouse X chromosome, making it the most extensively mapped mouse chromosome. The bulk of these probes are located in the central region of the mouse X chromosome and provide an extremely valuable framework for the physical mapping of the X-inactivation center region.

MAPPING THE MOUSE X-INACTIVATION CENTER

Genetic map in the vicinity of the mouse X-inactivation center

A large number of DNA markers, including microclones, linking clones, and genic probes, have been localized to the region of the mouse X-inactivation center. Analysis of backcross progeny from two interspecific crosses, one carrying the coat-texture mutations *Hq* and *Ta* (Brockdorff et al. 1987a) and the other carrying the *mdx* mutation (Cavanna et al. 1988), has identified a total of 12 mice with crossover events between the *Dmd* (*mdx*) and *Pgk-1* loci in the vicinity of the X-inactivation center. Pedigree analysis of a large number of linking clones, microclones, and genic probes in this panel of mice allowed the assignment and ordering of these clones to the X-inactivation region (Hamvas et al. 1990; Keer et al. 1990). To date, 20 probes have been mapped in detail to the region distal to *Dmd* (see Fig. 4) (N. Brockdorff et al., unpubl.). Several groups of clones appear to be inseparable by the available recombination events and are difficult to order with respect to each other. For example,

Table 1 Interspecific backcrosses employed for the genetic mapping of the mouse X chromosome

Source	Cross[a]	Segregating genetic markers[b]	Backcross progeny
Amar et al. (1985, 1988); Avner et al. (1987a)	B6CBARI-Ta ip/+ × Spe/Pas[c] backcrossed to B6CBARI- + +	Ta (Tabby), ip (jimpy) (also scored for $hprt$ protein polymorphism)	>200
Brockdorff et al. (1987a,b, 1988)	Hq Ta +/+ Li × $M.$ $spretus$ backcrossed to 129/J	Ta (Tabby) Hq (Harlequin)	234
Cavanna et al. (1988)	C57BL/10–mdx/mdx × $M.$ $spretus$ backcrossed to C57BL/10–mdx/mdx	mdx (X-linked muscular dystrophy)	90
Herman and Walton (1990)	B6CBA-$A$$^{w-J}$/$A$ BPA × $M.$ $spretus$ backcrossed to (C57BL/6J $A$$^{w-j}$ × CBA)F$_1$	Bpa (Bare patches)	22
Mullins et al. (1988, 1990)	C57BL/6 × $M.$ $spretus$ backcrossed to C57BL/6	none	100
Ryder-Cooke et al. (1988)	C57BL/10–mdx/mdx × $M.$ $spretus$ backcrossed to C57BL/10–mdx/mdx	mdx (X-linked muscular dystrophy)	>200

[a]The parental female laboratory mouse strain crossed to male $M.$ $spretus$ is indicated, and the male laboratory mouse strain to which F$_1$ female progeny were backcrossed is given.
[b]Visible phenotypic markers segregating in the backcross.
[c]Spe/Pas is an inbred $M.$ $spretus$ strain from the Institut Pasteur, Paris.

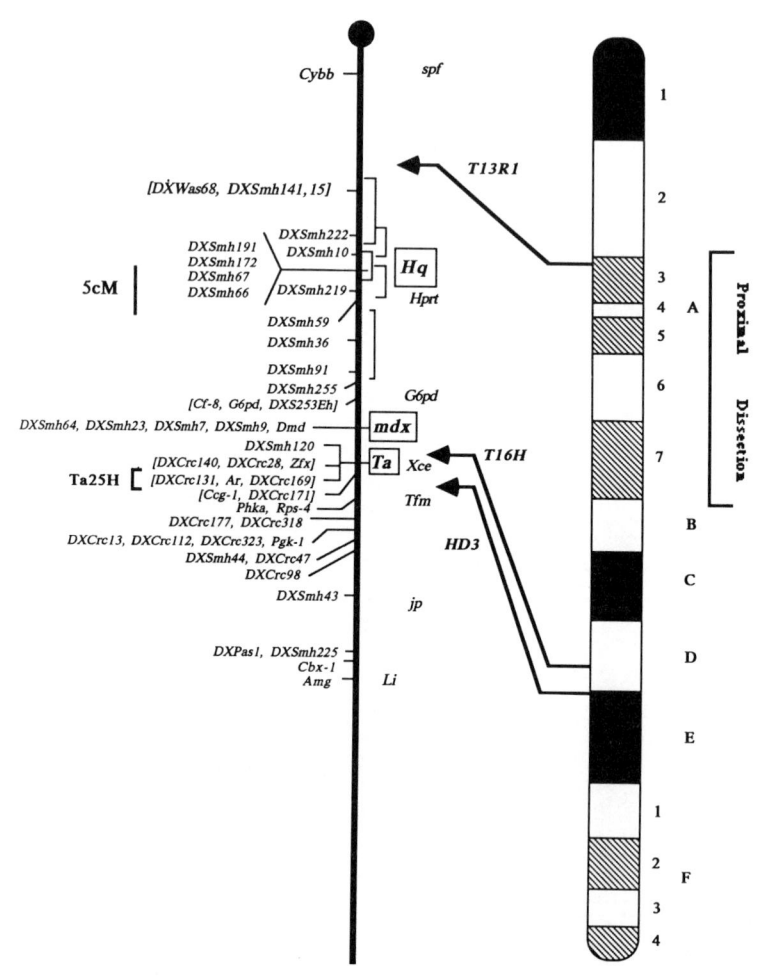

Figure 3 Molecular map of the mouse X chromosome. Genetic map positions of microclones, genic probes, and linking clones on the mouse X chromosome are illustrated to the left of the genetic map. Approximate limits of a regional microdissection of the mouse X chromosome (Fisher et al. 1985), are indicated (Proximal Dissection). Neighboring clones that fail to show recombination events are bracketed to the right of the map. Relative positions of the *Hq*, *Ta*, and *mdx* loci used in the interspecific backcrosses (Brockdorff et al. 1987a,b; Cavanna et al. 1988) are in boxes to the right of the map. Groups of loci encompassed by square brackets have been physically linked by PFGE. Concordance between the physical G-banded map of the chromosome and the genetic map is also indicated in three instances by the translocations T13R1, T16H, and the breakpoint HD3 (bold arrows).

MAPPING THE MOUSE X-INACTIVATION CENTER

Progeny / Loci	1723.4d	1724.3b	m13	m21	m17	1724.4f	1725.4e	m6	1722.4f	1722.6c	1722.7h	1722.6b
Dmd	D		S	D	SD			S	S	D		D
DXSmh120	SD	SD	S	D	SD	S	D	S	S	D	S	
DXCrc140	SD	SD	S	D	SD	S	D	S	S	D	S	SD
DXCrc28	SD	SD	S	D	SD	S	D	S	S	D	S	SD
Zfx	SD		S	D	SD		D	S	S	D	S	SD
DXCrc131	SD	SD	S	D	SD	S	D	S	S	D	S	SD
Ar	SD	SD	S	D	SD	S	D	S	S	D	S	SD
DXCrc169	SD	SD	S	D	SD	S	D	S	S	D	S	SD
Ta	SD	SD			S							
Ccg-1		D	D	D		S	D	S	S	D	S	SD
DXCrc171		D	D	D	SD	S	D	S	S	D	S	SD
Rps4x		D	D	S	D	S	D	S	S	D	S	SD
Phka		D	D	S	D	S	D	S	S	D	S	SD
DXCrc177		D	D	S	D	D	SD	S	S	D	S	SD
DXCrc318		D	D	S	D	D	SD	S	S	D	S	SD
Xist	SD	D	D	S	D	D	SD	S	S	D	S	SD
DXCrc112		D	D	S	D	D	SD	D	S	D	S	SD
DXCrc13		D	D	S		D	SD	D	S	D	S	
DXCrc323		D	D	S	D	D	SD	D	S	D	S	SD
Pgk-1		D	D	S	D	D	SD	D	S	D	S	SD
DXCrc47		D	D	S	D	D	SD	D	D	SD	D	SD
DXSmh44	SD	D	D	S	D	D	SD	D	D	SD	D	
DXCrc98		D	D	S	D	D	SD	D	D	SD	D	D

T16H, XIC, Ta 25H, XIC

C, D, E — T16H, X-inactivation Centre, HD3

▯ PHYSICAL LINK-UP | Ta25h: DELETION   XIC: X-INACTIVATION CENTRE

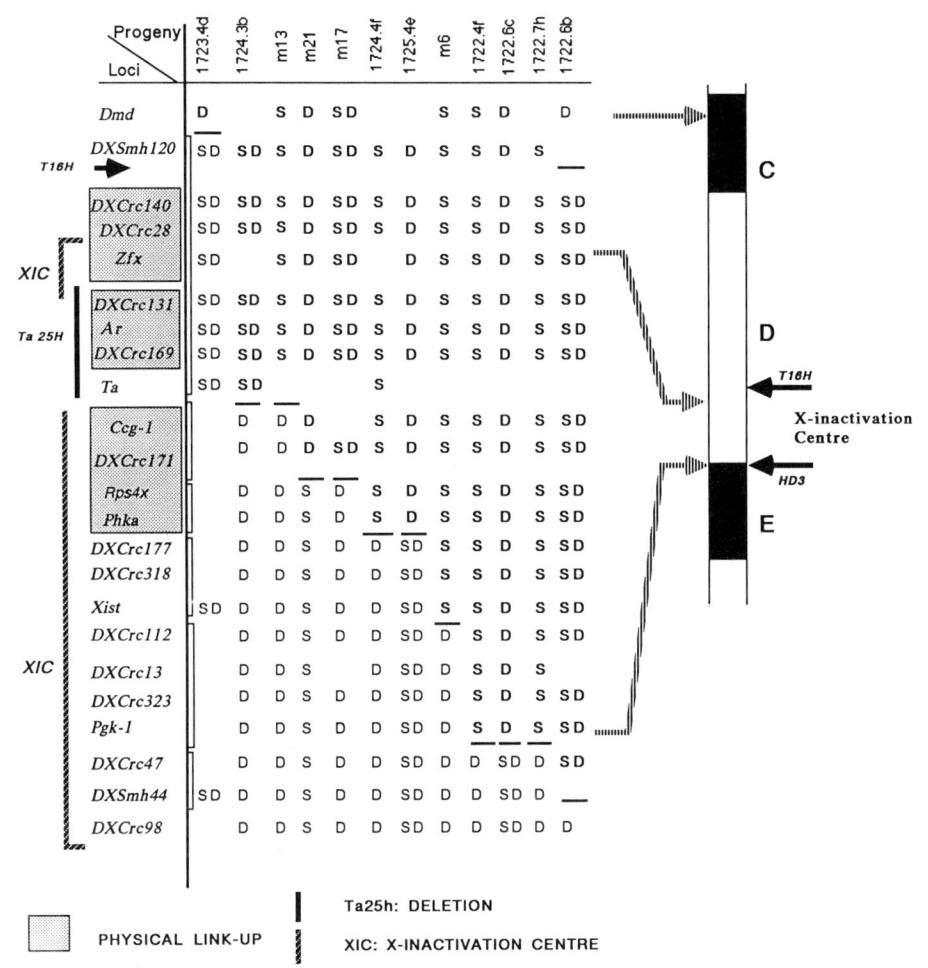

Figure 4 Molecular genetic analysis and mapping of clones in the vicinity of the mouse X-inactivation center. The table summarizes the *M. spretus* (S) and *M. domesticus* (D) scorings in selected mice carrying recombination breakpoints in the vicinity of the mouse X-inactivation center and allows the construction of a genetically ordered array of clones through this region (see text). Clones that fail to show recombination events cannot be ordered and are bracketed. In addition, the determined location of the T16H breakpoint and the extent of the Ta^{25H} deletion are indicated. The two regions of the X chromosome that may contain the X-inactivation center are also delineated (bold arrows). The chromosomal locations of the *Dmd*, *Zfx*, and *Pgk-1* loci as determined by in situ chromosomal hybridization are indicated. Note that the relative positions of the *Pgk-1* locus and the HD3 breakpoint are not determined. Clones that have been physically linked by PFGE are grouped in shaded boxes.

Mitchell et al. (1989) determined that *Zfx/Ar* mapped proximal to *Ta*, whereas Derry et al. (1989) reported the separation of *Zfx* from *Ar*, placing *Ar* distal to *Zfx*. Nevertheless, the inability to separate many of these clones testifies to their close linkage and to the possibility of linking them physically on the X chromosome map (see below).

In addition to genetic mapping, available DNA markers were also orientated with respect to the T16H breakpoint, the most proximal limit of the X-inactivation center, and the Ta^{25H} deletion, which was demonstrated not to contain X-inactivating sequences. Analysis of the microclone *DXSmh120* and the genic probes *Zfx* and *Ar* in the somatic cell hybrid B20c12 (Avner et al. 1987b), which contains only mouse X chromosome sequences proximal to the T16H translocation, indicated that the T16H breakpoint lies distal to *DXSmh120* but proximal to *Zfx* and *Ar* (Fig. 4) (Keer et al. 1990). As discussed above, the *Ar* and *Ta* loci are deleted in Ta^{25H} (Cattanach et al. 1989b). Further analysis of male mice carrying the Ta^{25H} deletion demonstrated that the *Eag*I linking clones *DXCrc131* (Hamvas et al. 1990) and *DXCrc169* are deleted in Ta^{25H} (N. Brockdorff et al., unpubl.), whereas the *Zfx* locus is unaffected. This is in agreement with genetic data that place *Zfx* proximal to the *Ar* locus (Derry et al. 1989). No information is available at this time on the location of the HD3 breakpoint on the genetic map.

Location of the mouse X-inactivation center with respect to the human X-inactivation center

It is pertinent at this juncture to reconsider the cytogenetic information on the location of the X-inactivation center in both the mouse and human X chromosome and how it relates to the extensive mouse genetic map in the region. It would appear from the available data that there are two potential regions for the location of the X-inactivation center in the mouse, flanking the Ta^{25H} deletion. Comparative mapping data would indicate that the inactivation center lies in the *Ar* to *Pgk-1* linkage group conserved between mice and humans. The proximal limit of this conserved segment on the mouse X chromosome lies somewhere between the *Ar* and *Zfx* loci, as evidenced by their widely separate positions on the human X chromosome (see Fig. 2). It is also important to note that *Zfx* and *Ar* lie distal to the T16H breakpoint (the proximal cytogenetic limit of the X-inactivation center) and that *Ar* lies within the Ta^{25H} deletion and *Zfx* does not. Therefore, the mouse X-inactivation center may lie in the small region proximal to the Ta^{25H} breakpoint and distal to *Zfx*.

Alternatively, the mouse X-inactivation center may lie distal to the Ta^{25H} deletion. In humans, evidence from the X;14 translocation (Brown et al. 1989) indicates that the X-inactivation center lies distal to *CCG1* and in the vicinity of *PGK1*. In mice, this region lies distal to the Ta^{25H} deletion and has been extensively mapped. Assuming that no smaller undiscovered rearrangements have occurred between mice and humans in the *Ar* to *Pgk-1* conserved linkage group, the region distal to the mouse *Ccg-1* locus appears to be the most likely location for the mouse X-inactivation center. Without further information on the location of the HD3 breakpoint in mouse, the most fruitful approach to the delineation of the distal limit of the mouse inactivation center is likely to be comparative analyses of conserved mouse clones in the various somatic cell hybrids that contain human X-autosome translocations delineating the distal limit of the human X-inactivation center.

Physical mapping of the mouse X-inactivation center region

The high density of available clones in the region of the mouse X-inactivation center (~20 clones distributed over 8 cM [~15 Mb]) augurs well for the construction of a physical long-range restriction map of the region. Groups of clones that lie within the same genetic interval and fail to demonstrate recombinants can be tested for physical linkage. The large restriction fragments of 50 kb to 10 Mb in size that are generated by rare-cutter restriction enzymes can be separated by pulsed-field gel electrophoresis (PFGE) (Barlow 1989) and probed with the available clones. Identification of commonly hybridizing pulsed-field fragments forms the first step in the physical linkage of adjacent clones and the construction of an encompassing pulsed-field restriction map. These technologies have already been used to demonstrate the physical linkage of the *G6pd*, *P3* (*DXS253Eh*), and *Cf-8* (Factor VIII) loci on the mouse X chromosome, which are encompassed within a 1.5-Mb long-range restriction map (Brockdorff et al. 1989).

Three groups of clones from the region of the mouse X-inactivation center have shown physical linkage when analyzed by PFGE (see Fig. 4). In one group of clones (*Zfx*, *DXCrc28*, and *DXCrc140*), a detailed restriction map has been constructed that encompasses approximately 1.5 Mb (R.M.J. Hamvas et al., unpubl.). Two other clones (*DXCrc202* and *DXCrc57*) that mapped to the same genetic interval were shown to be independently derived linking clones spanning the same CpG island as *DXCrc28*. This 1.5-Mb region must lie close to or within the putative X-inactivation region that lies proximal to Ta^{25H}. However, it has so far proved impossible to detect the proximal breakpoint of the Ta^{25H} deletion in the vicinity of this cluster of probes, so the orientation of the physical map is undetermined.

Clone contigs covering the X-inactivation region

Despite the success in achieving physical linkage of closely linked probes within the X-inactivation region, gaps in the map will inevitably occur. In addition, the PFGE linkage maps do not give access to the underlying sequences. Nevertheless, they provide the framework for coverage of the region with yeast artificial chromosome (YAC) clones. Such clones can carry large genomic inserts hundreds of kilobases in size (Burke et al. 1987) and will provide the resources for closing the gaps in the PFGE maps and presenting a complete clone contig covering the relevant regions.

A number of libraries carrying mouse partial *Eco*RI genomic inserts of large size are currently in preparation (Burke et al. 1991). The first stage in covering the PFGE map of the X-inactivation region with a clone contig will be to screen mouse YAC libraries with the available probes from the physical map. Fingerprinting of the isolated YAC clones by hybridization to mouse repeat sequences will allow identification of overlapping clones and the generation of contigs. Where gaps that prevent full coverage of the available physical map exist in the contig, end clones can be isolated by "bubble" PCR from the appropriate YACs that are terminal to the contigs (Riley et al., 1990). These end clones can be used to rescreen the YAC library in order to recover new clones that complete contig coverage. Furthermore, this process of end-clone recovery and rescreening of YAC libraries can be used as the basis for chromosomal walking between regions of the X-inactivation center that have not been linked by physical mapping and for which DNA markers are not available. In this way, an uninterrupted physical map of the X-inactivation center region for which all underlying sequences are available in YAC clones can be obtained. The physical size of the region distal to *Ccg-1* likely to contain the mouse X-inactivation center is still unknown. As indicated above, the distal limits of this region are most likely to be determined by comparative mapping of the distal limit of the human X-inactivation center on the mouse X chromosome. The region to be mapped and covered may be as big as 5 Mb. Nevertheless, the development of YAC technology has made the coverage and mapping of such an extensive chromosome region possible.

IDENTIFICATION OF THE MOUSE X-INACTIVATION CENTER

An X-inactivation assay

Ultimately, identification of the underlying sequences responsible for the initiation of X-inactivation will require the development of an assay for X-inactivation activity. This is particularly important since the relevant sequences may not be readily identifiable from sequence organization alone. The detailed nature of the X-inactivation process is an enigma, and the nature of the sequences involved is unknown. For example, the X-inactivation center may not encode a protein product, but may be composed of a protein recognition sequence or sequences or, alternatively, encompass a sequence region responsible for interaction with cellular components such as nuclear membrane attachment sites. If this is the case, direct sequence analysis will not necessarily uncover the relevant inactivation sequences. Sequences must be biologically tested for X-inactivation activity to identify those candidates that should undergo further detailed sequence analysis. Such an X-inactivation assay system has been proposed for the mouse (Rastan and Brown 1990).

A potential inactivation assay system in mice would be one employing ES cells. As indicated above, undifferentiated ES cells have not undergone X-inactivation; following differentiation, female ES cells inactivate one or the other of their two X chromosomes. If an X-inactivation center were introduced into a male undifferentiated ES cell, it would then possess two X-inactivation centers. Following differentiation, there would be two possible outcomes: (1) the X-inactivation center residing on the single X chromosome would be inactivated or (2) the introduced X-inactivation center would undergo inactivation. Either one of these possibilities forms the basis of testing sequences for X-inactivation activity.

A proposed application of this procedure involves E14TG2a ES cells. These cells carry a deletion of the X-linked hypoxanthine phosphoribosyltransferase (*Hprt*) gene (Thompson et al. 1989) and are sensitive to hypoxanthine aminopterin thymidine (HAT) (Fig. 5). Constructs of candidate X-inactivation sequences, carrying a neomycin (*neo*) selectable marker and a functional *Hprt* minigene, could be introduced into E14TG2a cells. Stable integrants of the construct would be selected by HAT. Subsequently, selected integrants would be allowed to undergo differentiation and X-inactivation, followed by selection with 6-thioguanine (6-TG). 6-TG selects against cells carrying an active *Hprt* gene. If a functional X-inactivation center has been introduced on the construct, then approximately 50% of the cells will inactivate the normal X chromosome (presumably lethal in a male cell), whereas the remaining cells will inactivate at the integrated construct, thus inactivating the neighboring functional *Hprt* minigene. These cells will survive the 6-TG selection and around half the cell population will remain viable. However, if a functional inactivation sequence has not been introduced on the construct, none of the cells will be protected from back selection by 6-TG on the functional *Hprt* minigene. By introducing a *neo* marker on the construct, the inactivation of a gene near the putative functional X-inactivation center that is not part of the selectable system can also be confirmed; surviving cells would be expected to be *neo*-sensitive following differentiation and 6-TG selection.

It is important to note that the presence of a functional X-inactivation center should result in 50% cell survival after selection with 6-TG. This result is demonstrably different from recovering a few colonies of cells that may arise from low levels of excision and elimina-

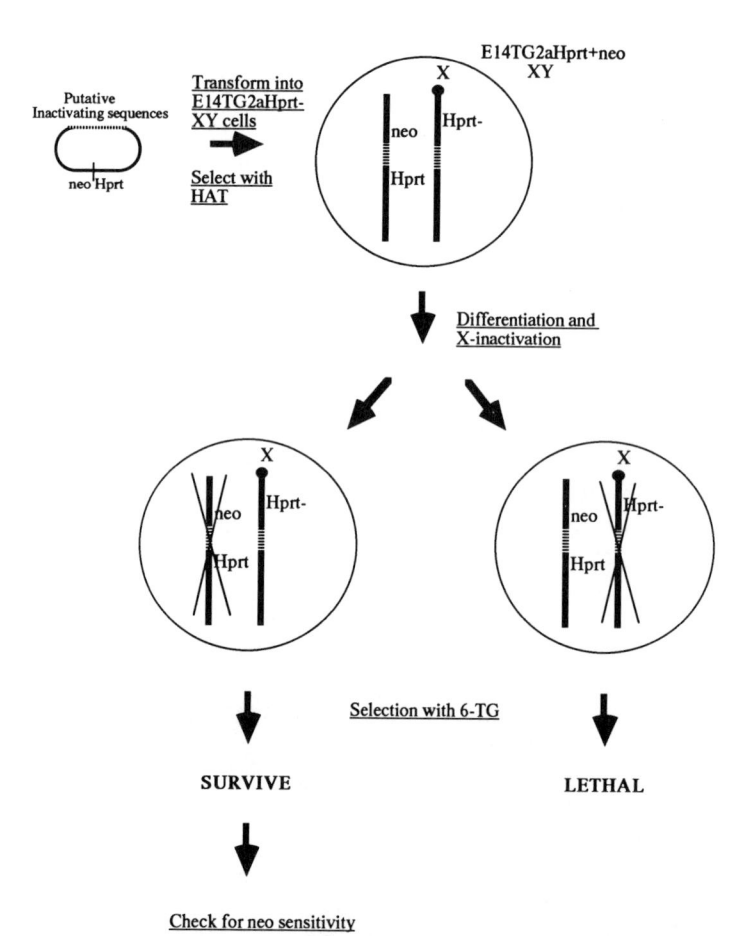

Figure 5 Experimental strategy for a functional X-inactivation assay.

tion of the integrated construct and functional *Hprt* minigene. In addition, the autosomal monosomy that may result from inactivation of autosomal regions neighboring the integrated construct should not be lethal to cells growing in vitro. Nevertheless, this objection can be met by testing a number of different integrants of each potential inactivation sequence.

Finally, with no knowledge of the underlying organization or extent of the inactivation center sequences, it may be necessary to screen YAC constructs with the X-inactivation assay to demonstrate a functional inactivation center activity. However, the stable integration of YACs into ES cell genomes has yet to be demonstrated. Nevertheless, this assay may provide a route to examine the function of a human cDNA (*XIST*) that maps to the X-inactivation region and that is expressed only on the inactivated X chromosome.

CONCLUSION

Molecular and genetic mapping is facilitating the accurate delineation of the X-inactivation center region of the mouse X chromosome. A physical map of the X-inactivation region can provide the framework for the construction of a YAC contig that covers the relevant region and provides access to all of the underlying sequences. Relevant human or mouse sequences that map to the X-inactivation region may be tested for functional X-inactivation center characteristics by using an X-inactivation assay. In this way, the various genetic interactions or pathways to the developmental switch of X-inactivation can hopefully be identified.

Acknowledgments

I thank my colleagues Jacquie Keer, Neil Brockdorff, Renata Hamvas, Sohaila Rastan, and Mary Lyon, all of whom have contributed greatly to the work and ideas expressed in this paper. This work was supported by grants G822373CB and G8803031CB from the Medical Research Council, United Kingdom.

References

Allderdice, P.W., O.J. Miller, D.A. Miller, and H.P. Klinger. 1978. Spreading of inactivation in an (X;14) translocation. *Am. J. Med. Genet.* **2:** 233.

Amar, L.C., D. Arnaud, J. Cambrou, J.L. Guenet, and P.R. Avner. 1985. Mapping of the mouse X chromosome using random genomic probes and an interspecific mouse cross. *EMBO J.* **4:** 3695.

Amar, L.C., L. Dandalo, A. Hanauer, A. Ryder-Cook, D. Arnaud, J.L. Mandel, and P. Avner. 1988. Conservation and reorganisation of loci on the mammalian X chromosome: A molecular framework for the identification of homologous subchromosomal regions in man and mouse. *Genomics* **2:** 220.

Avner, P., L. Amar, D. Arnaud, A. Hanauer, and J. Cambrou. 1987a. Detailed ordering of markers localising to the Xq26-Xpter region of the human chromosome by the use of an interspecific *Mus spretus* mouse cross. *Proc. Natl. Acad. Sci.* **84:** 1629.

Avner, P., D. Arnaud, L. Amar, J. Cambrou, H. Winking, and L.B. Russell. 1987b. Characterisation of a panel of somatic cell hybrids for regional mapping of the mouse X chromosome. *Proc. Natl. Acad. Sci.* **84:** 5330.

Barlow, D.P. 1989. Pulsed-field gel electrophoresis. *Genome* **31:** 465.

Brockdorff, N., L.C. Amar, and S.D.M. Brown. 1989. Pulse-field linkage of the P3, G6pd, and Cf8 genes on the mouse X chromosome: Demonstration of synteny at the physical level. *Nucleic Acids Res.* **17:** 1315.

Brockdorff, N., M. Montague, S. Smith, and S. Rastan. 1990. Construction and analysis of the linking libraries from the mouse X chromosome. *Genomics* **7:** 573.

Brockdorff, N., E.M.C. Fisher, J.S. Cavanna, M.F. Lyon, and S.D.M. Brown. 1987a. Construction of a detailed molecular map of the mouse X chromosome by microcloning and interspecific crosses. *EMBO J.* **6:** 3291.

Brockdorff, N., E.M.C. Fisher, S.H. Orkin, M.F. Lyon, and S.D.M. Brown. 1988. Localisation of the human X-linked gene for chronic granulomatous disease to the mouse X chromosome. *Cytogenet. Cell Genet.* **48:** 124.

Brockdorff, N., G.S. Cross, J.S. Cavanna, E.M.C. Fisher, M.F. Lyon, K.E. Davies, and S.D.M. Brown. 1987b. The mapping of a cDNA from the human X-linked Duchenne muscular dystrophy gene to the mouse X chromosome. *Nature* **328:** 166.

Brown, C.J. and H.F. Willard. 1989a. Localisation of a gene that escapes inactivation to the X chromosome proximal short arm: Implications for X inactivation. *Am. J. Hum. Genet.* **45:** 592.

———. 1989b. Localisation of the X inactivation centre (XIC) to Xq13. *Cytogenet. Cell Genet.* **51:** 971.

Brown, C.J., A.M. Flenniken, B.R.G. Williams, and H.F. Willard. 1990. X chromosome inactivation of the human TIMP gene. *Nucleic Acids Res.* **18:** 4191.

Brown, C.J., T. Sekiguchi, T. Nishomoto, and H.F. Willard. 1989. Regional localisation of the CCF1 gene which complements hamster cell cycle mutation BN462 to Xq11-Xq13. *Somatic Cell Mol. Genet.* **15:** 93.

Brown, C.J., A. Ballabio, J.L. Rupert, R.G. Lafreniere, M. Grompe, R. Tonlorenzi, and H.F. Willard. 1991a. A gene from the region of the human X inactivation centre is expressed exclusively from the inactive X chromosome. *Nature* **349:** 38.

Brown, C.J., R.G. Lafreniere, V.E. Powers, G. Sebastio, A. Ballabio, A.L. Pettigrew, D.H. Ledbetter, E. Levy, I.W. Craig, and H.F. Willard. 1991b. Localisation of the X inactivation centre on the human X chromosome in Xq13. *Nature* **349:** 82.

Brown, S.D.M. 1985. Mapping mammalian chromosomes: New technologies for old problems. *Trends Genet.* **1:** 219.

———. 1991. Physical mapping of the mouse genome. In *Techniques in the behavioural and neural sciences. Techniques for the genetic analysis of brain and behaviours: Focus on the mouse* (ed. D. Goldowitz et al.). Elsevier, New York. (In press.)

Brown, S.D.M. and A.J. Greenfield. 1987. A model to describe the size distribution of mammalian genomic fragments recovered by microcloning. *Gene* **55:** 327.

Brown, S.D.M., N. Brockdorff, J.S. Cavanna, E.M.C. Fisher, A.J. Greenfield, M.F. Lyon, and J. Nasir. 1988. The long-range mapping of mammalian chromosomes. *Curr. Top. Microbiol. Immunol.* **137:** 3.

Burke, D.T., G.F. Carle, and M.V. Olson. 1987. Cloning of large segments of exogenous DNA into yeast by means of artificial chromosome vectors. *Science* **236:** 806.

Burke, D.T., J.M. Rossi, J. Leung, D.S. Koos, and S.M. Tilghman. 1991. A mouse genomic library of yeast artificial chromosome clones. *Mamm. Genome* **1:** 65.

Cattanach, B.M. 1966. The location of Cattanach's translocation into the X-chromosome linkage map of the mouse. *Genetical Res.* **8:** 253.

———. 1974. Position effect variegation in the mouse. *Genetical Res.* **23:** 291.

Cattanach, B.M., C. Raspberry, and S.J. Andrews. 1989a. Further *Xce* linkage data. *Mouse News Lett.* **83**: 165.

Cattanach, B.M., C. Raspberry, E.P. Evans, and M. Burtenshaw. 1989b. Ta^{25H}, a presumptive X chromosome deletion. *Mouse News Lett.* **83**: 160.

Cattanach, B.M., C. Raspberry, E.P. Evans, L. Dandalo, M.C. Simmler, and P. Avner. 1991. Genetic and molecular evidence of a X chromosome deletion spanning the tabby (*Ta*) and testicular feminisation (*Tfm*) loci in the mouse. *Cytogenet. Cell Genet.* (in press).

Cavanna, J.S., G. Coulton, J. Morgan, N. Brockdorff, S.M. Forrest, K.E. Davies, and S.D.M. Brown. 1988. Molecular and genetic mapping of the mouse *mdx* locus. *Genomics* **3**: 337.

Ceci, J.D., L.D. Siracusa, N.A. Jenkins, and N.G. Copeland. 1989. A molecular genetic linkage map of mouse chromosome 4 including the localisation of several proto-oncogenes. *Genomics* **5**: 699.

Derry, J.M.J., A.S. Ryder-Cooke, D.B. Lubahn, F.S. French, E.M. Wilson, D.C. Page, and P.J. Barnard. 1989. Localisation of the androgen receptor gene on the mouse X chromosome and its relationship to X-linked zinc finger protein. *Cytogenet. Cell Genet.* **51**: 988.

Disteche, C.M., G.K. McConnell, S.G. Grant, D.A. Stephenson, V.M. Chapman, S. Gandy, and D.A. Adler. 1989. Comparison of the physical and recombination maps of the mouse X chromosome. *Genomics* **5**: 177.

Disteche, C.M., L.M. Kunkel, A. Lojeswki, S.H. Orkin, M. Eisenhard, E. Sahar, B. Travis, and S.A. Latt. 1982. Isolation of mouse X-chromosome specific DNA from an X-enriched lambda phage library derived from flow-sorted chromosomes. *Cytometry* **2**: 282.

Eicher, E.M. 1970. X-autosome translocations in the mouse: Total inactivation versus partial inactivation of the X chromosome. *Adv. Genet.* **16**: 175.

Fisher, E.M.C., J.S. Cavanna, and S.D.M. Brown. 1985. Microdissection and microcloning of the mouse X chromosome. *Proc. Natl. Acad. Sci.* **82**: 5486.

Fisher, E.M.C., P. Beer-Romero, L.G. Brown, A. Ridley, J.A. McNeil, J.B. Lawrence, H.F. Willard, F.R. Bieber, and D.C. Page. 1990. Homologous ribosomal protein genes on the human X and Y chromosomes: Escape from X inactivation and implications for Turner Syndrome. *Cell* **63**: 1205.

Goodfellow, P., B. Pym, T. Mohandas, and L.J. Shapiro. 1984. The cell surface antigen locus, MIC2X, escapes X-inactivation. *Am. J. Hum. Genet.* **36**: 777.

Graves, J.A.M. and S.M. Gartler. 1986. Mammalian X chromosome inactivation: Testing the hypothesis of transcriptional control. *Somat. Cell Mol. Genet.* **12**: 275.

Hamvas, R.M.J., J.T. Keer, N. Brockdorff, and S.D.M. Brown. 1990. The genetic mapping of an Eag I linking clone, EM131, to the *Zfx/Ar* cluster of the mouse X chromosome. *Mouse Genome* **87**: 114.

Herman, G.E. and S.J. Walton. 1990. Close linkage of the murine locus Bare Patches to the X-linked visual pigment gene: Implications for mapping human X-linked dominant Chondrodysplasia Punctata. *Genomics* **7**: 307.

Johnston, P.G. and B.M. Cattanach. 1981. Controlling elements in the mouse. IV. Evidence of non-random X-inactivation. *Genetical Res.* **37**: 151.

Keer, J.T., R.M.J. Hamvas, N. Brockdorff, D. Page, S. Rastan, and S.D.M. Brown. 1990. Genetic mapping in the region of the mouse X-inactivation center. *Genomics* **7**: 566.

Kratzler, P.G. and V.M. Chapman. 1981. X chromosome reactivation in oocytes

of *Mus caroli*. *Proc. Natl. Acad. Sci.* **78:** 3093.
Kratzer, P.G. and S.M. Gartler. 1978. HGPRT activity changes in preimplantation mouse embryos. *Nature* **274:** 503.
Lindsay, S. and A.P. Bird. 1987. Use of restriction enzymes to detect potential gene sequences in mammalian DNA. *Nature* **327:** 336.
Lubahn, D.B., D.R. Joseph, P.M. Sullivan, H.F. Willard, F.S. French, and E.M. Wilson. 1988. Cloning of the human androgen receptor complementary DNA and localisation to the X chromosome. *Science* **240:** 327.
Ludecke, H.-J., G. Senger, U. Claussen, and B. Horsthemke. 1989. Cloning defined regions of the human genome by microdissection of banded chromosomes and enzymatic amplification. *Nature* **338:** 348.
Lyon, M.F. 1961. Gene action in the X chromosome of the mouse (*Mus musculus* L.). *Nature* **190:** 373.
———. 1988. X-chromosome inactivation and the location and expression of X-linked genes. *Am. J. Hum. Genet.* **42:** 8.
Lyon, M.F. and A.G. Searle. 1989. *Genetic variants and strains of the laboratory mouse*, 2nd edition. Oxford University Press, England.
Lyon, M.F., J. Zenthon, E.P. Evans, M.D. Burtenshaw, K.A. Wareham, and E.D. Williams. 1986. Lack of inactivation of a mouse X-linked gene physically separated from the inactivation centre. *J. Embryol. Exp. Morphol.* **97:** 75.
Mattei, M.G., J.F. Mattei, I. Vidal, and F. Giraud. 1981. Structural anomalies of the X chromosome and inactivation centre. *Hum. Genet.* **56:** 401.
McBurney, M.W. 1988. X chromosome inactivation: A hypothesis. *Bioessays* **9:** 85.
Mitchell, M., D. Simon, N. Affara, M. Ferguson-Smith, P. Avner, and C. Bishop. 1989. Localisation of murine X and autosomal sequences homologous to the human Y located testis-determining region. *Genetics* **121:** 803.
Mohandas, T., R.L. Geller, P.H. Yen, J. Rosendorff, R. Bernstein, A. Yoshida, and L.J. Shapiro. 1987. Cytogenetic and molecular studies on a recombinant human X chromosome: Implications for the spreading of X chromosome inactivation. *Proc. Natl. Acad. Sci.* **84:** 4954.
Monk, M. and M.I. Harper. 1979. Sequential X chromosome inactivation coupled with cellular differentiation in early mouse embryos. *Nature* **281:** 311.
Mullins, L.J., S.G. Grant, D.A. Stephenson, and V.M. Chapman. 1988. Multilocus molecular map of the mouse X chromosome. *Genomics* **3:** 187.
———. 1990. Efficient linkage of 10 loci in the proximal region of the mouse X chromosome. *Genomics* **7:** 19.
Nadon, N., N. Dorn, and R. DeMars. 1988. A-11: Cell type-specific and single-active-X transcription controls of newly found gene in cultured human cells. *Somat. Cell Mol. Genet.* **14:** 541.
Poutska, A. and H. Lehrach. 1986. Jumping libraries and linking libraries: The next generation of molecular tools in mammalian genetics. *Trends Genet.* **2:** 174.
Rastan, S. 1983. Non-random X-chromosome inactivation in mouse X-autosome translocation embryos—Location of the inactivation centre. *J. Embryol. Exp. Morphol.* **78:** 1.
Rastan, S. and S.D.M. Brown. 1990. The search for the mouse X-chromosome inactivation centre. *Genetical Res.* **56:** 99.
Rastan, S. and E.J. Robertson. 1985. X-chromosome deletions in embryo-derived

(EK) cell lines associated with lack of X-chromosome inactivation. *J. Embryol. Exp. Morphol.* **90:** 379.

Riley, J., R. Butler, D. Ogilvie, R. Finniear, D. Jenner, S. Powell, R. Anand, J.C. Smith, and A.F. Markham. 1990. A novel, rapid method for the isolation of terminal sequences from yeast artificial chromosome (YAC) clones. *Nucleic Acids Res.* **18:** 2887.

Roberts, B., P. Barton, A. Minty, P. Daubas, A. Weyder, F. Bonhomme, J. Catalan, D. Chazottes, J.L. Guenet, and M. Buckingham. 1985. Investigation of genetic linkage between myosin and actin genes using an interspecific mouse backcross. *Nature* **314:** 181.

Robertson, E.J., M.J. Evans, and M.H. Kaufman. 1983. X-chromosome instability in pluripotential stem cell lines derived from parthogenetic embyros. *J. Embryol. Exp. Morphol.* **74:** 297.

Russell, L.B. 1963. Mammalian X-chromosome action: Inactivation limited in spread and in region of origin. *Science* **140:** 976.

Russell, L.B. and N.L.A. Cacheiro. 1978. The use of mouse X-autosome translocations in the study of X-inactivation pathways and nonrandomness. In *Genetic mosaics and chimaeras in mammals* (ed. L.B. Russell), p. 393. Plenum Press, New York.

Russell, L.B. and C.S. Montgomery. 1970. Comparative studies on X-autosome translocations in the mouse. II. Inactivation of autosomal loci, segregation and mapping of autosomal breakpoints in five T(X;1)'s. *Genetics* **64:** 281.

Ryder-Cook, A.S., P. Sicinski, K. Thomas, K.E. Davies, R.G. Worton, E.A. Barnard, M.G. Darlison, and P.J. Barnard. 1988. Localisation of the *mdx* mutation within the mouse dystrophin gene. *EMBO J.* **7:** 3017.

Schmidt, M. and B. Migeon. 1990. Asynchronous replication of homologous loci on human active and inactive X chromosomes. *Proc. Natl. Acad. Sci.* **87:** 3685.

Schneider-Gadicke, A., P. Beer-Romeo, L.G. Brown, R. Nussbaum, and D.C. Page. 1989. Zfx has a gene structure similar to Zfy, the putative human sex determinant, and escapes X inactivation. *Cell* **57:** 1247.

Seldin, M.F., T.A. Howard, and P. D'Eustachio. 1989. Comparison of linkage maps of mouse chromosome 12 derived from laboratory strain intraspecific and *Mus spretus* interspecific backcrosses. *Genomics* **5:** 24.

Shapiro, L.J., T. Mohandas, R. Weiss, and G. Romero. 1979. Non-inactivation of an X-chromosome locus in man. *Science* **204:** 1224.

Smith, C.L., S.K. Lawrence, G.A. Gillespie, C.R. Cantor, S.M. Weissman, and F.S. Collins. 1987. Strategies for mapping and cloning macroregions of mammalian genomes. *Methods Enzymol.* **151:** 461.

Tabor, A., O. Andersen, E. Niebuhr, and H. Sardemann. 1983. Interstitial deletion in the critical region of the long arm of the X chromosome in a mentally retarded boy and his normal mother. *Hum. Genet.* **64:** 196.

Tagaki, N. and G.R. Martin. 1984. Studies of the temporal relationship between the cytogenetic and biochemical manifestations of X-chromosome inactivation during the differentiation of LT-1 and teratocarcinoma stem cells. *Dev. Biol.* **103:** 425.

Thompson, S., A.R. Clarke, A.M. Pow, M.L. Hooper, and D.W. Melton. 1989. Germ line transmission and expression of a corrected HPRT gene produced by gene targeting in embryonic stem cells. *Cell* **56:** 313.

The Regulation of the Human β-Globin Locus

Niall Dillon, Dale Talbot, Sjaak Philipsen, Olivia Hanscombe, Peter Fraser, Sara Pruzina, Michael Lindenbaum, and Frank Grosveld

Laboratory of Gene Structure and Expression
National Institute for Medical Research
The Ridgeway, Mill Hill, London NW7 1AA, England

The aim of this chapter is to provide a review of the regulation of the human β-globin locus.

The main topics discussed include:

❑ a summary of the locus control region and the DNA-binding factors that interact with this region

❑ the mechanisms of integration-position-independent and copy-number-dependent gene expression

❑ the genetics of the human β-globin gene cluster with respect to developmental expression

❑ transgenic mice as models for studying human β-globin regulation

❑ a novel model to explain the developmental regulation of the human β-globin gene locus

INTRODUCTION

The human β-globin gene cluster spans a region of 70 kb containing five developmentally regulated genes in the order $5'$-ε, γ_G, γ_A, δ, β-$3'$. In the early stages of human development, the embryonic yolk sac is the hematopoietic tissue and expresses the ε-globin gene. This is followed by a switch to the γ-globin genes in the fetal liver and the δ- and β-globin genes in adult bone marrow (for review, see Collins and Weissman 1984). These genes are expressed at exceptionally high levels, giving rise to 90% of the total soluble protein in circulating red blood cells. These cells are derived from a pluripotent hematopoietic stem cell, which can self-renew or differentiate along alternate pathways to erythrocytes, platelets, granulocytes, macrophages, and lymphocytes. Differentiation takes place via a series of intermediate "committed" precursors that progressively lose proliferative capacity and become more restricted in developmental potential. During the final transition to erythroblasts that have lost the capacity to proliferate, the β-globin genes become transcriptionally activated, achieving mRNA levels of more than 25,000 copies per cell.

The entire β-globin-like gene locus has been sequenced, and a large number of structural defects in and around the β-globin gene have been documented (for review, see Collins and Weissman 1984; Poncz et al. 1989). These defects are responsible for a heterogeneous group of genetic diseases collectively known as the β-thalassemias, which are classified into subgroups (e.g., β-, δβ-, and γδβ-thalassemia) according to the type of gene affected. In a related condition termed hereditary persistence of fetal hemoglobin (HPFH), γ-globin gene expression and fetal hemoglobin (HbF) production persist into adult life. These diseases are clinically important and provide natural models for the study of the regulation of globin gene transcription and the mechanism of gene switching during development. Most interesting in terms of transcriptional regulation are the promoter mutations and deletions. The δβ-thalassemias and a number of the HPFHs are the result of deletions of varying sizes and are associated with an elevated expression of the γ gene in adult life. Analysis of these deletions has suggested that they act over considerable distances to influence differential gene expression within the human β-globin domain.

THE LOCUS CONTROL REGION

The presence of a region that controls the activity of the entire β-globin gene cluster first became apparent from the study of a human γβ-thalassemia (Kioussis et al. 1983). The patient involved had one normal chromosome 11, but the other allele had undergone a 100-kb deletion that eliminated the entire upstream region, leaving the β-globin gene intact, but silent (Taramelli et al. 1986). Cloning and expressing the β gene from the mutant allele showed that it was completely normal (Kioussis et al. 1983; Wright et al. 1984). The normal allele was expressed in the patient, indicating that it was not a lack of *trans*-acting factors that was responsible for the silencing of the mutant chromosome, but rather that

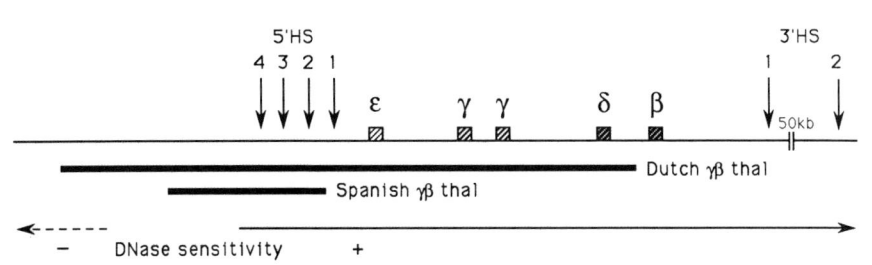

Figure 1 Schematic representation of the β-globin locus. Boxes indicate the transcriptional orientation of the different genes from left to right. Four vertical arrows upstream of the ε gene mark the LCR containing the DNase I hypersensitive sites 5'HS1, HS2, HS3, and HS4. Two vertical arrows downstream of the β gene mark the 3'HS1 and HS2. Black bars under the locus represent two thalassemic conditions (Dutch γβ, Spanish γβ) in which in vivo deletions eliminate the function of the LCR. Horizontal arrows indicate low (–) and high (+) sensitivity of the chromatin to DNase I digestion.

an important control region had been deleted. Analysis of the upstream region of the ε-globin gene showed the presence of a set of developmentally stable sites that displayed hypersensitivity to the nuclease DNase I. These HS sites, labeled 5'HS1, HS2, HS3, and HS4, were potential candidates for such a locus control region (LCR) (Fig. 1) (Tuan et al. 1985; Forrester et al. 1987; Grosveld et al. 1987). Linkage of this region to a cloned β-globin gene resulted in erythroid-specific, high-level expression of the gene in transgenic mice and in stably transformed tissue culture cells. This expression was independent of the integration site in the host genome and was dependent on the copy number of the transgene (Grosveld et al. 1987; Blom van Assendelft et al. 1989), a phenomenon that had not been observed previously in transgene expression. This posed two questions: (1) How is independence of the site of integration achieved? (2) How is activation of β-globin expression by the LCR regulated at the molecular level? The answer to these fundamental questions is not yet known, but some progress has been made.

POSITION INDEPENDENCE

Position independence and copy number dependence can theoretically be explained by at least two independent mechanisms: Either positive activation by the LCR is always achieved and can result in very high levels of expression that obscure small position effects that may still be present in the background, or (and) the region contains sequences that insulate it from neighboring regions, providing a border to functional domains.

Border sequnces

A border function could possibly be fulfilled by matrix attachment sites (MAR) (Gasser and Laemmli 1986). We initially speculated that both activating sequences and potential locus border elements (LBE) were part of the LCR (Grosveld et al. 1987). This is based on the fact that the DNase I sensitivity of chromatin in isolated nuclei is strongly decreased in the sequences 25–30 kb upstream of the LCR (see Fig. 1) (Kioussis et al. 1983; Taramelli et al. 1986; Forrester et al. 1990). One strong MAR just upstream of the LCR was present on our original constructs and a potential candidate for a border function (Jarman and Higgs 1988). However, several other MARs were mapped at the same time in the downstream region, which cannot fulfill such a role. In the downstream direction, the chromatin remains sensitive under the control of the LCR for at least 150 kb (Forrester et al. 1990), suggesting that no such sequences are present around the 3'HS1 (Fig. 1). Possibly, the upstream MAR was an "A" element (a subset of MAR elements), which is thought to be responsible for the position-independent expression of the chicken lysozyme gene in cultured cells and transgenic mice (Stief et al. 1989; Bonifer et al. 1990). However, our preliminary data using the 5'MAR of the globin locus indicate that it does not have "insulating" properties, in agreement with the fact that it is located in a DNase-I-sensitive area (D. Greaves, unpubl.). It is not yet clear whether such functional elements are present further upstream of the β-globin locus, coincident with the change of DNase I sensitivity.

Dominant activation

It appears that the position independence that we observe is, at least in part, due to the fact that the LCR achieves activation of transcription in some dominant fashion, perhaps by creating very stable interactions between the LCR and the gene. Consequently, positive position effects would only be present as part of the background and would only become apparent in situations where the linked gene is suppressed (see below for discussion) (Dillon and Grosveld 1991). Interestingly, position effects are not observed when low levels of expression are obtained by the use of part of the LCR or mutations in the LCR, indicating that the interaction between the (part) LCR and the promoter is dominant except when the promoter is suppressed (Forrester et al. 1989; Ryan et al. 1989; Talbot et al. 1989; Collis et al. 1990; Fraser et al. 1990; Dillon and Grosveld 1991). It is clear that the main activity is associated with HS2, HS3, and HS4 (Forrester et al. 1989; Ryan et al. 1989; Talbot et al. 1989; Tuan et al. 1989; Collis et al. 1990; Fraser et al. 1990), in agreement with the deltion observed in a Hispanic γβ-thalassemia (Driscoll et al. 1989). Moreover, each of these sites can confer copy-number-dependent expression

Figure 2 Factor binding sites on the minimal fragment of 5′HS2, HS3, and HS4, providing position-independent expression in transgenic mice. Individual factors are described in the text. Black boxes (NF-E2, GATA-1, Jun) indicate erythroid-specific factors. Open boxes (H-BP, J-BP) indicate ubiquitous factors. Restriction sites and GT-rich base pair motifs (GT) (Philipsen et al. 1990) are indicated above the boxes. Arrows indicate orientation of the binding sites.

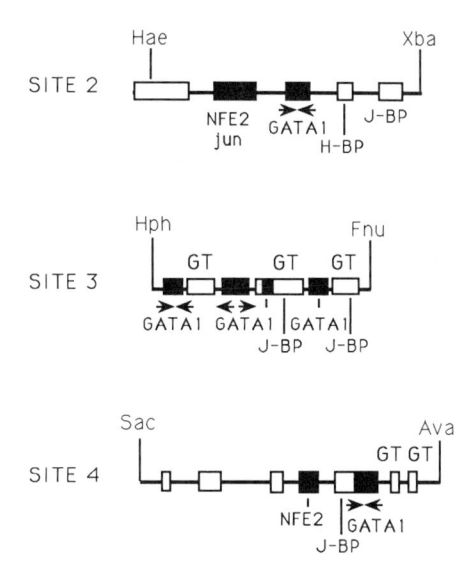

on the β-globin gene in transgenic mice, indicating that each of the sites contains an element capable of activating a linked transgene, independent of the site of integration.

Factors and binding sites

Transient transfection experiments show that classical enhancer activity is only associated with the 5′HS2 (Tuan et al. 1989; Ney et al. 1990) and not with the other HS sites. Dissection of the HS2 showed that a number of proteins are bound to the core fragment (Fig. 2) (Talbot et al. 1990). Attention has been focused on a double consensus sequence for the Jun/Fos family of DNA-binding proteins, which appeared to be crucial for HS2 activity (Ney et al. 1990; Sorrentino et al. 1990; Talbot et al. 1990). Several lines of evidence show that the functional activator which interacts with the Jun/Fos-binding site is NF-E2, originally described as a factor that interacts with the promoter of another erythroid-specific gene, the porphobilinogen deaminase (PBGD) gene (Mignotte et al. 1989a). Upon erythroid differentiation of MEL cells, the level of Jun/Fos proteins is reduced relative to that of NF-E2 (Mignotte et al. 1989a; Talbot et al. 1990). Point mutagenesis of the G residues flanking the consensus sequence leaves Jun/Fos binding intact but reduces the binding of NF-E2 (Mignotte et al. 1989b; Talbot et al. 1990). This was confirmed by Ney et al. (1990), who also show that these mutations result in reduced activity of this enhancer element alone or when linked to the 5′HS1. Our recent experiments show that the protein interactions at this site are complicated, involving two NF-E2 molecules and at least one other protein binding at two nonequivalent sites (Talbot and Grosveld 1991). However, the presence of this double NF-E2 sequence alone is insufficient to provide high levels of expression of a linked β-globin gene in transgenic mice (Talbot et al. 1990). Moreover, when the Jun/Fos-binding site is removed from the 300-bp core fragment, HS2 retains the ability to activate a linked β-globin gene in a copy-number-dependent fashion, albeit at low levels (Talbot and Grosveld 1991). We therefore conclude that the 5'HS2 NF-E2 region has strong enhancer activity and is important for obtaining high levels of globin gene expression but that it is not involved or required to obtain position-independent globin gene activation.

Five other major protein-binding sites are present within the HS2 core fragment (see Fig. 2). Three of these sites bind ubiquitously expressed nuclear factors; the other two bind (at least in part) an erythroid-specific component (Talbot et al. 1990). It is interesting to note that a number of these binding sites are also present in 5'HS3 and 5'HS4, the other two active sites (Fig. 2) (Philipsen et al. 1990; Pruzina et al. 1991). One of the shared factors is the erythroid/mega-karyocyte-specific factor GATA-1 (Martin and Orkin 1990; Romeo et al. 1990), which has been shown to be essential for erythroid development (Pevny et al. 1991). Deletion of GATA-1-binding sites prevents erythroid-specific induction of the β-globin gene (de Boer et al. 1988), and the protein has been shown to have transcriptional activation properties (Martin and Orkin 1990). GATA-1 binds to many erythroid-specific genes, including all of the globin genes, but the presence of GATA-1-binding sites per se is insufficient to give position-independent expression. The immediate 5'- and 3'-flanking regions of the human β-globin gene contain at least six GATA-1-binding sites (de Boer et al. 1988; Wall et al. 1988), but they do not confer integration-site-independent expression on the β-globin gene (Magram et al. 1985; Townes et al. 1985; Kollias et al. 1986). However, all 5'HS sites contain two closely spaced GATA-1 sites in opposite orientation. This arrangement is also observed in the chicken β-globin enhancer. A chicken β-globin gene containing the enhancer has recently been shown to be expressed in a position-independent manner (Reitman et al. 1990). It is possible that an inverted double GATA site is a key component in erythroid-specific, position-independent activation and that GATA-1 can interact with itself or one of the other GATA proteins to achieve this activation (Yamamoto et al. 1990).

All of the other factors that have been shown to interact with LCR sequences, including the factors H-BP and J-BP, are ubiquitous proteins (Talbot and Grosveld 1991), suggesting that a combination of erythroid-specific and ubiquitous factors may be required to render the β-globin gene independent of its site of integration. Two abundant ubiquitous factors shared by the three HS sites of the LCR which have been studied to date are Sp1 and TEF-2 (Gidoni et al. 1985; Xiao et al. 1987), but a simple multimerized combination of GATA-1- and Sp1/TEF-2-binding sites is not functional (S. Philipsen, unpubl.). It is therefore conceivable that other as yet less well characterized factors are involved in LCR function. These factors may not be abundant and could be overlooked in DNase I footprinting and bandshift experiments with crude nuclear extracts. We are currently fractionating erythroid extracts to address this question.

DEVELOPMENTAL REGULATION OF THE β-GLOBIN LOCUS

The LCR is required for high-level expression of the β, γ, and ε genes (Grosveld et al. 1987; Behringer et al. 1990; Shih et al. 1990; Dillon and Grosveld 1991), which raises the question of how the genes are activated separately at different stages of development. There are two types of model which are not mutually exclusive that could explain this process. In the first model, called the local control model, the LCR provides an activating function at all stages of development, but its action on individual genes at particular developmental stages is blocked by the binding of stage-specific negative factors to individual gene promoters. In other words, the developmental control of the genes is specified by sequences immediately flanking the genes. This model predicts the presence of binding sites for stage-specific transcriptional activators and/or suppressors close to the gene (see below).

The second model, called the competition model, envisages competition between the genes for direct interaction with the LCR as a major determinant of stage specificity. According to this model, the LCR can activate any of the genes at all stages, but it is prevented from doing so by a dominant interaction with a particular gene or genes at each stage. Predictions are more difficult to make for this model because they strongly depend on what is understood by "competition." For example, competition for a limited amount of a particular factor between binding sites in different promoters is very different from a competition between two blocks of regulatory sequences for interaction with a third block that depends on the presence or absence of particular factors in one of the blocks. Competition could be influenced by active transcription or by spatial constraints, such as the position of the genes within the domain.

Deletions and point mutations

Evaluation of these models is assisted by the existence of a number of deletions and point mutations in the locus that are associated with persistence of γ-gene expression into adult life (Fig. 3). Point mutations in the γ promoters that have been linked to HPFH phenotypes can be

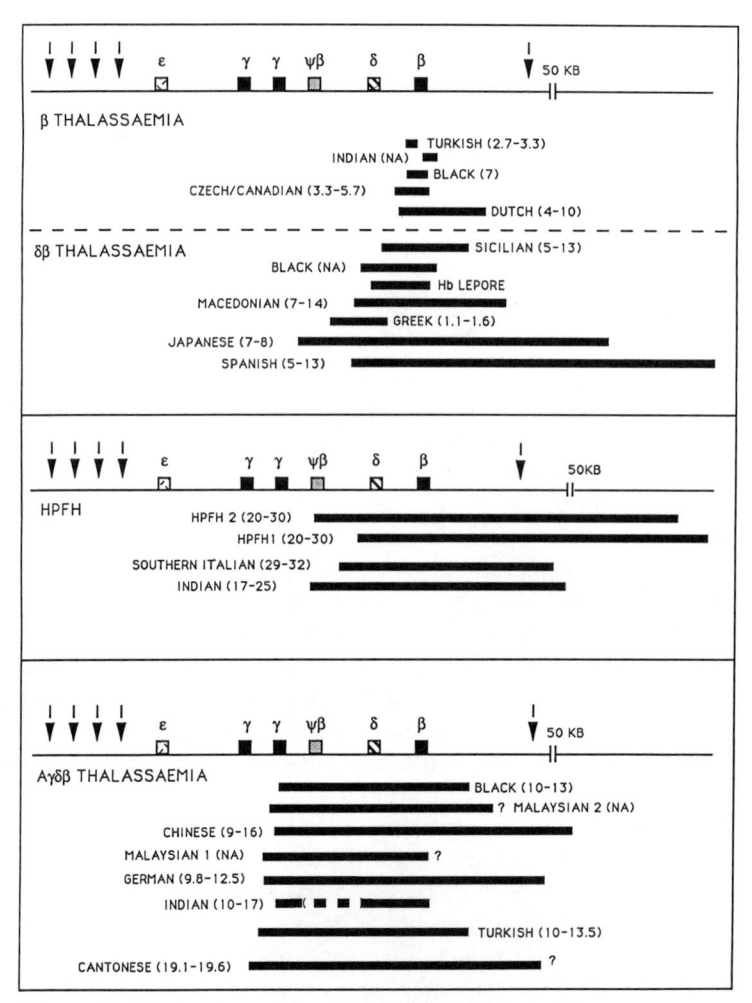

Figure 3 Schematic representation of the different deletions occurring in the β-globin locus in thalassemias and HPFHs. Arrows indicate the position of the 5′ and 3′HS sites. The position of each globin gene is indicated as a shaded box. Black bars indicate the size of the deletion, and numbers in brackets indicate the levels of γ-globin expression in heterozygotes.

divided into two groups. Mutations that lie upstream of base −150 result in the creation of new or improved binding sites for the transcription factors Sp1 (Gumucio et al. 1988; Fischer and Nowock 1990) and GATA-1 (Mantovani et al. 1988; Martin et al. 1989). The second group of mutations are clustered around the distal CAAT box and appear to result in the loss of factor binding sites (Mantovani et al. 1989; Fucharoen et al. 1990), suggesting that this region may contain a binding site for a negative regulator such as NF-E3.

An increase in adult γ-gene expression in HPFH is associated with a corresponding down-regulation of the β gene on the same chromosome (Weatherall and Clegg 1981; Giglioni et al. 1984). This down-regulation and the fact that increased expression of the γ genes in adult life is associated with some deletions that involve the β gene might at first sight appear to support a model for stage-specific regulation by reciprocal competition between the genes (Blom van Assendelft et al. 1989). However, a closer examination suggests that although the γ gene is directly involved in the down-regulation of the β gene, the β gene is not involved in the silencing of the γ gene in adult life. Down-regulation of the β gene in the nondeletion HPFHs is dependent on active γ-gene transcription resulting from single point mutations in the γ promoters. The reduction in β expression to about 60% is approximately equivalent to the rise in γ-gene expression, with only a slight reduction in overall transcriptional output from the locus (Giglioni et al. 1984). This suggests that any competition taking place is linked to transcription.

The existence of transcriptional competition in the locus is also suggested by the relative trancription levels from the $γ_A$ and $γ_G$ genes in the fetal liver (Huisman et al. 1977; Schroeder 1980). However, there is no evidence to suggest that this type of transcriptional competition plays any significant role in silencing the γ genes in adult life. Point mutations in the β promoter that drastically reduce the level of β transcription do not significantly increase γ-gene expression. Likewise, in the Turkish and Czech/Canadian β-thalassemias, where the β promoter has been deleted and β expression has been completely abolishied, γ expression levels average less than 5% (see Fig. 3) (Poncz et al. 1989).

Larger deletions (>10 kb) are associated with higher levels of γ-gene expression, but the levels and cellular distributions of γ chains vary widely between different deletions, making it unlikely that they are caused by a single mechanism. These deletions can be grouped into three categories: Aγδβ-thalassemias, deletion HPFHs, and δβ-thalassemias (see Fig. 3). The Aγδβ-thalassemias all have deletions that extend into the region of γ transcription, rendering them uninformative for competition models, because γ-gene silencer sequences may have been deleted. The high-level pancellular γ expression observed in the deletional HPFHs places them in a separate category; HPFH deletion breakpoints are clustered, which has led to the suggestion that sequences at the breakpoints could be responsible for the adult γ-gene expression. Indeed, enhancer-like sequences have been found close to several of the HPFH breakpoints (Feingold and Forget 1989; Anagnou et al. 1990). The remaining deletions giving rise to significant levels of γ-gene expression in the adult include the δβ-thalassemias and Dutch β-thalassemia (see Fig. 3). The level of γ chains in individuals heterozygous for these conditions shows a broad range, extending from 5% to 15% in individuals with the same deletion. It is necessary to be cautious about attributing these levels directly to a transcriptional effect caused by the deletion. A proportion of individuals heterozygous for nondeletion β^0-thalassemias already show elevated levels of γ-globin gene expression (up to 5%), indicating that chain imbalance can act to amplify the basal level of γ globin (Weatherall and Clegg 1981; Nathan 1990).

A further important feature of γ-gene expression in the δβ-thalassemias is the fact that it is largely restricted to a subset of erythrocytes (heterocellular expression). Acid elution and immunofluorescent staining both show that many cells contain no detectable γ globin (Weatherall and Clegg 1981), an observation that creates problems for loop exclusion competition models for γ silencing (Townes and Behringer 1990). The similarity of phenotypes between the δβ-thalassemias and Dutch β-thalassemia suggests that a minimum deletion size, rather than removal of the δ gene, is the parameter that gives rise to higher levels of γ expression. Sensitivity to DNase I digestion in erythroid cells has been shown to extend at least 150 kb downstream from the β gene (Forrester et al. 1990), and an erythroid-specific transcription unit has been identified in this downstream region. It is possible that the region downstream from the β gene insulates the γ gene from an erythroid-specific domain downstream from the locus. According to this scenario, large deletions involving this region expose the γ genes to weak activation by the downstream domain, whereas high-level pancellular expression characteristic of the HPFHs results from deletions that place the γ genes close to specific enhancer sequences within this domain. The possibility that the spatial distribution of the various elements within the locus may also play an important role in the normal situation will be considered further below, where the results from transgenic mice studies are discussed.

Transgenic mice systems

Attempts to resolve the questions raised above have made use of transgenic mice as a model system for studying human globin gene switching. Mice do not possess separate fetal globin genes but instead switch directly from embryonic to adult β-globin expression at 11–13 days of gestation.

ε-Gene regulation The developmental regulation of the human ε gene has been analyzed in embryonic stem (ES) cells and transgenic mice. In the absence of the LCR, the ε gene is not expressed in mice (Shih et al. 1990). When linked to the LCR, ε is expressed at a high level during the embryonic stage and is completely silenced in the fetal liver and adult bone marrow (Lindenbaum and Grosveld 1990; Raich et al. 1990; Shih et al. 1990; P. Watt et al., in prep.). Expression studies with deletion mutants in K562 cells suggest that the region of –200 to –300 in the ε-globin promoter might play a role in silencing the ε gene (Cao et al. 1989). When deletion mutants lacking this sequence were analyzed in transgenic mice, a small increase in ε-gene expression in adult mice was observed, but the level remained low, indicating that other sequences are probably also involved in ε-gene silencing (P. Watt et al., in prep.). A candidate DNA-binding protein for this suppressor function has been identified. It binds to this region and is stage-specific but not erythroid-specific (P. Watt et al., in prep.). The position of other suppressor-binding sequences are presently not known, but a potential site is the CAAT box region in the promoter that shows good homology with the distal CAAT box region of the γ gene (see below).

γ-*Gene regulation* In the absence of the LCR, the human γ transgene is expressed during the embryonic stage in mice and is switched off with the mouse embryonic genes (Chada et al. 1986; Kollias et al. 1986). Initial reports suggested that linkage of a γ gene to the LCR resulted in γ-gene expression at all developmental stages and that the γ gene was silenced in adult mice when the β gene was present downstream from the γ gene. This appeared to support a competition model, where the presence of the β gene is required for silencing of the γ gene (Behringer et al. 1990; Enver et al. 1990). However, when the single γ-gene experiment was carried out on animals with only one or two copies of the LCR–γ-gene construct, a different result was obtained. Although γ-gene expression persisted in the early fetal liver, it was silenced at adult stages, independent of the presence of the β gene (Dillon and Grosveld 1991). Interestingly, this silencing can be perturbed by the position of integration of the transgene in the mouse genome (Dillon and Grosveld 1991). This important result indicates that, like the ε gene, transcription of the γ gene in the presence of the LCR can be completely blocked by the action of stage-specific negative regulators and removes the basis of the argument that the β gene would be needed for γ silencing. The result also demonstrates that the silencing can be disrupted by neighboring sequences brought into close proximity to the γ gene, an observation that fits well with the genetic data from large deletions discussed above.

The elements responsible for γ-gene silencing have not yet been identified, but the mutations associated with the nondeletion HPFHs suggest that the sequences around the distal CAAT box are likely to be involved. For example, a 13-bp deletion that removes the distal CAAT box results in a very strong HPFH (60% γ-gene expression level) (Gilman et al. 1988). Interestingly, a recently described Japanese HPFH (20% expression level) is associated with a point mutation in the CAAT sequence of the distal CAAT box (Fucharoen et al. 1990), which reduces affinity for the transcription factor CP1 (Chodosh et al. 1988). The base –117 mutation associated with Greek HPFH (40% expression level) has been reported to cause reduced binding of the erythroid-specific factor NF-E3

(Mantovani et al. 1989). These findings suggest a model for γ-gene silencing in which factors binding to the distal CAAT box (at base −115) compete for interaction with factors bound to upstream promoter sequences, thereby preventing the proximal CAAT box (at base −87) from forming such interactions. The distal CAAT box is located outside the normal optimal position for CAAT elements, and this is likely to prevent it from functioning as an effective positive promoter element. One would expect this type of silencing mechanism to be very dependent on both the topology of the promoter region and the creation of extra factor binding sites in the upstream sequences. Such binding sites may partially bypass or change the competition between the proximal and distal CAAT boxes, resulting in suboptimal transcription levels. The availability of a transgenic mouse model for γ-gene silencing should allow these ideas to be tested.

β-Gene regulation Linkage of the adult β gene to the LCR results in inappropriate expression at the embryonic stage (Blom van Assendelft et al. 1989; Behringer et al. 1990; Enver et al. 1990; Lindenbaum and Grosveld 1990; Hanscombe et al. 1991), albeit at a lower level than that of the mouse embryonic genes. This expression can be blocked by placing a γ gene or a human α-globin gene between the β gene and the LCR (Behringer et al. 1990; Enver et al. 1990; Hanscombe et al. 1991), supporting the idea that competition plays a role in preventing premature β expression. However, when the order is reversed and the β gene is placed in the first position, it is expressed at a level similar to that observed for the β gene in the absence of the γ or α gene (Fig. 4) (Hanscombe et al. 1991). This indicates that silencing of the β gene at the embryonic stage is not caused by simple competition between the genes but that the relative distance (i.e., position) between the LCR and the genes is important for their precise developmental expression pattern. This effect appears to be a function of both the nonequivalence of the promoters and their position relative to each other and to the LCR (see below).

Polarity of the β-globin locus has long been suggested by the fact that the genes are arranged in the order of their expression during development, and the discovery of the LCR 5′ to the ε gene adds a further dimension. The order of the genes is conserved among mammals, but there is some divergence in the other vertebrate loci that have been characterized. In chicken, the embryonic ε and ρ genes are located at opposite ends of the locus, with the adult β genes between them. However, it is important to note that the chicken ε gene is a minor embryonic gene that contributes only 20% of the total embryonic Hb, compared to 80% for ρ (Brown and Ingram 1974). In addition, there are indications that the chicken β-globin LCR may have been split such that part of it is located between the β and ε genes (Reitman et al. 1990), perhaps as a result of an ancient translocation event that placed the ε gene in its pres-

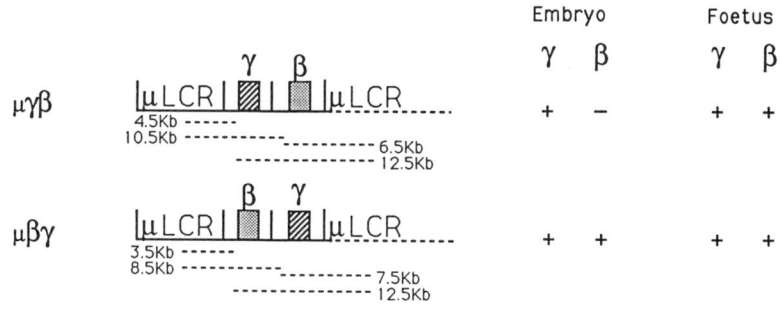

		Embryo		Foetus	
		γ	β	γ	β
μγβ	μLCR [γ] [β] μLCR 4.5Kb 10.5Kb 6.5Kb 12.5Kb	+	−	+	+
μβγ	μLCR [β] [γ] μLCR 3.5Kb 8.5Kb 7.5Kb 12.5Kb	+	+	+	+

Figure 4 Micro-locus (μLCR) γβ and βγ constructs (Hanscombe et al. 1991). Genes (shaded boxes) are in the same 5' to 3' transcriptional orientation with respect to each other and the LCR. Dotted LCR lines illustrate the distance from a promoter to a 5' and 3' LCR in multicopy animals. Expression of the γ and β genes in fetal and embryonic transgenic mice is indicated by + and − symbols.

ent position in the chicken locus. A model for regulation of the chicken genes through competition was first suggested by a study on the ε and β genes (Choi and Engel 1988), but it is so far based solely on transient expression assays and needs to be validated further in a transgenic system.

MECHANISM OF GLOBIN GENE SWITCHING

The data reviewed above indicate that developmental regulation of the human β-globin locus is a complex process with several different modes of regulation. An overall pattern does emerge that centers around the polarity of the locus. The earliest gene to be activated, the ε gene, is also the one in closest proximity to the LCR. It is possible that the γ genes (and the β gene) are suppressed by competition with the ε gene, although there are no experimental data to support this. Alternatively or in addition, the γ and β genes may bind embryonic stage-specific factors that keep their promoters suppressed. The silencing of the ε gene in the fetal liver by the binding of one or more suppressor factors to the ε promoter negates its competitive ability, the γ genes are expressed, and they in turn keep expression of the β gene suppressed by competition. The γ genes are switched off during the period around birth, again by stage-specific negative regulators acting on sequences in the promoter to silence the γ genes. As a consequence, the β gene is activated and expressed in the bone marrow.

How could the expression of the upstream genes prevent premature expression of those located downstream, but not vice versa, when it is clear from the transgenic data that there are big differences in the efficiency of the ε- and γ-, but not the β-, gene promoters at the different developmental stages? The mechanism by which the proximal genes can suppress those located more distally is unknown, but we suggest an interesting possibility involving loop formation and relative distance to the LCR. The data in favor of loop formation versus scanning models between enhancers and promoters are well established. Of particular importance in this regard are the experiments which have shown that even non-DNA mediated linkage of an enhancer to a promoter is functional (Muller et al. 1989) and those on transvection in Drosophila (Bickel and Pirotta 1990 and references therein). Distance has also been established as an important parameter using multiple-linked genes in different systems, such as the nonerythroid transient assays (de Villiers et al. 1983; Wasylyk et al. 1983) or the transgenic mice discussed above (Hanscombe et al. 1991).

We propose that loop formation between regulatory elements and the frequency of interaction between the promoters and the LCR are dependent on the effective volume in which these elements operate. This effect would be most pronounced if the LCR and the genes were all present on one structural chromatin loop of a size that is several times the distance between the LCR and the genes. Structural chromatin loops of more than 100 kb are observed in mammalian cell nuclei (Gasser and Laemmli 1986; Cockerhill and Garrard 1986 and references therein), and the fact that the LCR controls DNase I hypersensitivity of the β-globin locus over at least 150 kb (Forrester et al. 1990) suggests that the entire β-globin locus may be present on one very large chromatin loop. If we assume this to be the case, the frequency of interactions between any of the promoters with the LCR would be proportional to their effective concentration relative to the LCR. On the basis of ring closure probabilities with naked DNA, the effective concentration of two points on the DNA will be related to the volume of a sphere and will be proportional to the power of 3/2 of the distance. This concept is illustrated in Figure 5. Applying the rule to the β locus, the β gene is twice as far as the γ_G gene from the HS2 enhancer of the LCR. Therefore, the β gene occupies an approximately threefold larger volume relative to the HS2 enhancer of the LCR than the γ_G gene, which should give it an eightfold lower frequency of interaction with the LCR (see Fig. 5).

This effect will work in favor of the proximal gene decreasing the affinity differences required for competition, but it will work against the distal gene. Distal genes would be incapable of suppressing upstream genes under similar circumstances unless the downstream gene promoter increased its affinity by several orders of magnitude relative to the upstream gene. The transgenic mouse data on the expression of the β-globin gene at the embryonic and fetal/adult stages argue strongly against this possibility. Instead, the problem is solved by local suppression of the upstream promoters to allow expression from the downstream gene (Fig. 6). This model predicts that a repeat of the gene order experiment illustrated in Figure 4 incorporating a very substantial in-

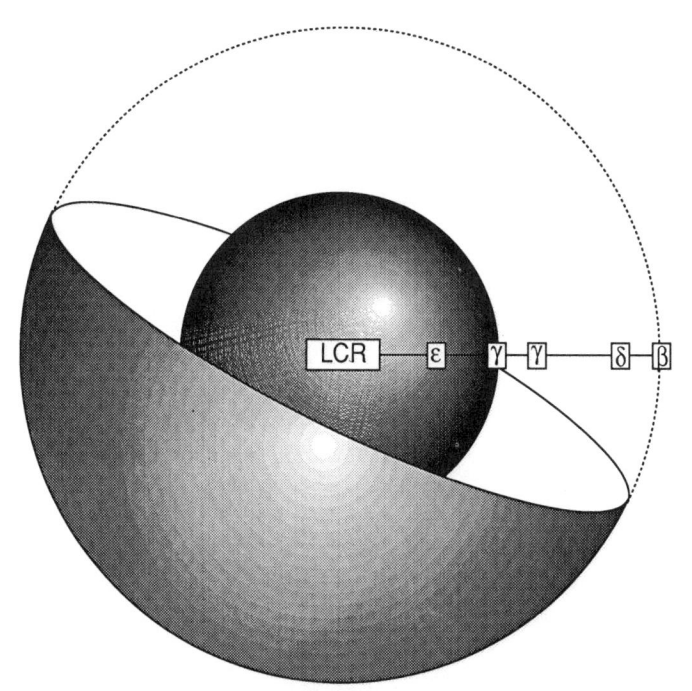

Figure 5 Schematic representation of the relative volumes occupied by the γ_G and β genes relative to the LCR. For simplicity of presentation, the LCR is shown as a fixed point in the center of the sphere. Only half of the β-globin gene outer sphere is shown.

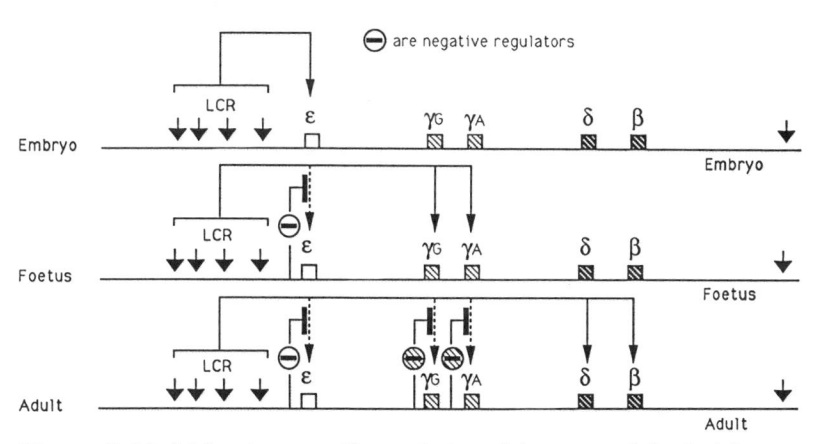

Figure 6 Model for stage-specific regulation of the genes of the β-globin locus. (*Short arrows*) 5' and 3' HS sites; (*solid arrows*) activation of genes by the LCR; (*dashed arrows*) blocked activation. Stage-specific negative factors silencing the gene are illustrated by a bar inside a circle. The location of these is not accurate and there may be more than one factor for each gene.

crease in the distance between the LCR and the genes should yield a different result for the relative expression of the γ and β genes at the different developmental stages. Experiments to substantiate or disprove this prediction are presently in progress.

Finally, the LCR may exert different effects on different genes in the locus. The transgenic mouse data already show that the ε gene is more dependent on the LCR for expression than are the other genes, and there are indications that each gene may have evolved a promoter that precisely matches its position in the locus relative to the LCR. The factors involved in stage-specific regulation of the ε, γ, and β genes are not yet characterized. Factors such as GATA-1 or NF-E2 are present at all stages of development, although changes in their concentrations may be important (Whitelaw et al. 1990). The current information available on the ε gene suggests that although such positively acting factors determine the erythroid specificity of the β-globin locus, developmental regulation may be mediated by factors that are stage-specific but not erythroid-specific. Such factors may have been recruited in an ad hoc manner as the genes evolved and very different factors may be used for similar globin genes of different species.

References

Anagnou, N.P., C. Perez-Stable, R. Gelinas, F. Costantini, K. Liapaki, M. Constantopoulou, T. Costeas, N. Moschonas, and G. Stamatoyannopoulos. 1990. DNA sequences residing 3' of the breakpoint of the HPFH-3 deletion can modify the developmental regulation of the fetal Aγ globin gene. *Clin. Res.* **38:** 301A.

Behringer, R.R., T.M. Ryan, R.D. Palmiter, R.L. Brinster, and T.M. Townes. 1990. Human γ to β globin gene switching in transgenic mice. *Genes Dev.* **4:** 380.

Blom van Assendelft, G., O. Hanscombe, F. Grosveld, and D.R. Greaves. 1989. The β-globin domain control region activates homologous and heterologous promoters in a tissue-specific manner. *Cell* **56:** 969.

Bickel, S. and V. Pirotta. 1990. Self association of the *Drosophila* zeste protein is responsible for transvection effects. *EMBO J.* **9:** 2959.

Bonifer, C., M. Vidal, F. Grosveld, and A.E. Sippel. 1990. Tissue specific and position independent expression of the complete gene domain for chicken lysozyme in transgenic mice. *EMBO J.* **9:** 2843.

Brown, J. and V. Ingram. 1974. Structural studies on chick embryonic hemoglobins. *J. Biol. Chem.* **249:** 3960.

Cao, S., P.D. Gutman, H.P.G. Dave, and A.J. Schechter. 1989. Identification of a transcriptional silencer in the 5'-flanking region of the human ε-globin gene. *Proc. Natl. Acad. Sci.* **86:** 5306.

Chada, K., J. Magram, and F. Costantini. 1986. An embryonic pattern of expression of a human fetal globin gene in transgenic mice. *Nature* **319:** 685.

Chodosh, C., A. Baldwin, R. Canthew, and P. Sharp. 1988. Human CCAAT-

binding proteins have heterologous sub-units. *Cell* **53**: 11.
Choi, O.R. and J.D. Engel. 1988. Developmental regulation of human β globin gene switching. *Cell* **55**: 17.
Cockerhill, P. and W. Garrard. 1986. Chromosomal loop anchorage of the kappa immunoglobulin gene occurs next to the enhancer in a region containing topoisomerase II sites. *Cell* **44**: 273.
Collins, F.S. and S.M. Weissman. 1984. The molecular genetics of human hemoglobin. *Prog. Acid Res. Mol. Biol.* **31**: 315.
Collis, P., M. Antoniou, and F. Grosveld. 1990. Definition of the minimal requirements within the human β-globin gene and the dominant control region for high level expression. *EMBO J.* **9**: 233.
de Boer, E., M. Antoniou, V. Mignotte, L. Wall, and F. Grosveld. 1988. The human β-globin gene promoter; nuclear protein factors and erythroid specific induction of transcription. *EMBO J.* **7**: 4203.
de Villiers, J., C. Olson, J. Banerji, and W. Schaffner. 1983. Analysis of the transcriptional enhancer effect. *Cold Spring Harbor Symp. Quant. Biol.* **47**: 911.
Dillon, N. and F. Grosveld. 1991. Human γ-globin genes silenced independently of other genes in the β-globin locus. *Nature* **350**: 252.
Driscoll, C., C. Dobkin, and B. Alter. 1989. γδβ-Thalassemia due to a de novo mutation deleting the 5′ β-globin gene activation-region hypersensitive sites. *Proc. Natl. Acad. Sci.* **86**: 7470.
Enver, T., N. Raich, A.J. Ebens, T. Papayannopoulou, F. Costantini, and G. Stamatoyannopoulos. 1990. Developmental regulation of human fetal-to-adult globin gene switching in transgenic mice. *Nature* **344**: 309.
Feingold, E. and B. Forget. 1989. The breakpoint of a large deletion causing hereditary persistence of fetal hemoglobin occurs within an erythroid DNA domain remote from the β globin gene cluster. *Blood* **74**: 2178.
Fischer, K. and J. Nowock. 1990. The T to C substitution at ~198 of the Aγ globin gene associated with the British form of HPFH generates overlapping recognition sites for two DNA binding proteins. *Nucleic Acids Res.* **18**: 5685.
Forrester, W., U. Novak, R. Gelinas, and M. Groudine. 1989 Molecular analysis of the human β-globin locus activation region. *Proc. Natl. Acad. Sci.* **86**: 5439.
Forrester, W., S. Takegawa, T. Papayannopoulou, G. Stamatoyannopoulos, and M. Groudine. 1987. Evidence for a locus activator region. *Nucleic Acids Res.* **15**: 10159.
Forrester, W., E. Epner, C. Driscoll, T. Enver, M. Brice, T. Papayannopoulou, and M. Groudine. 1990. A deletion of the human β globin locus activation region causes a major alteration in chromatin structure and replication across the entire β globin locus. *Genes Dev.* **4**: 1637.
Fraser, P., J. Hurst, P. Collis, and F. Grosveld. 1990. DNase I hypersensitive sites 1, 2, and 3 of the human β-globin dominant control region direct position-independent expression. *Nucleic Acids Res.* **18**: 3503.
Fucharoen, S., K. Shimiza, and M. Fukumaki. 1990. A novel C-T transition within the distal CCAAT motif of the Gγ globin gene in the Japanese HPFH: Implication of factor binding in elevated fetal globin expression. *Nucleic Acids Res.* **18**: 5245.
Gasser, S. and U. Laemmli. 1986. Cohabitation of scaffold binding regions with upstream/enhancer elements of three developmentally regulated genes in

D. melano-gaster. Cell **46**: 521.

Gidoni, D., J.T. Kadonaga, H. Barrera-Saldana, K. Takahashi, P. Chambon, and R. Tjian. 1985. Bidirectional SV40 transcription mediated by tandem Sp1 binding interactions. *Science* **230**: 511.

Giglioni, B., C. Casini, R. Mantovani, S. Merli, P. Comp, S. Ottolenghi, G. Saglio, C. Camaschella, and U. Mazza. 1984. A molecular study of a family of Greek hereditary persistence of fetal hemoglobin and β-thalassaemia. *EMBO J.* **11**: 2641.

Gilman, J., N. Mishima, X. Wen, T. Stoming, J. Lobel, and T. Huisman. 1988. Distal CCAAT box deletion in the Aγ globin gene of two black adolescents with elevated fetal Aγ globin. *Nucleic Acids Res.* **18**: 10635.

Grosveld, F., G. Blom van Assendelft, D. Greaves, and G. Kollias. 1987. Position-independent high level expression of the human β-globin gene in transgenic mice. *Cell* **51**: 975.

Gumicio, D., K. Rood, T. Gray, M. Riordan, C. Sartor, and F. Collins. 1988. Nuclear proteins that bind the human γ globin gene promoter: Alterations in binding produced by point mutations associated with hereditary persistence of fetal hemoglobin. *Mol. Cell. Biol.* **8**: 5310.

Hanscombe, O., D. Whyatt, P. Fraser, N. Yannoutsos, D. Greaves, and F. Grosveld. 1991. Globin gene order is important for correct developmental expression. *Genes Dev.* (in press).

Huisman T., H. Harris, and M. Gravely. 1977. The chemical heterogeneity of the fetal hemoglobin in normal newborn infants and in adults. *Mol. Cell. Biochem.* **17**: 45.

Jarman, A. and D. Higgs. 1988. Nuclear scaffold attachment sites in the human globin gene complexes. *EMBO J.* **7**: 3337.

Kioussis, D., E. Vanin, T. de Lange, R.A. Flavell, and F. Grosveld. 1983. β-globin gene inactivation by DNA translocation in γ-thalassaemia. *Nature* **306**: 662.

Kollias, G., N. Wrighton, J. Hurst, and F. Grosveld. 1986. Regulated expression of Aγ-, β-, and hybrid γβ-globin genes in transgenic mice: Manipulation of the developmental expression pattern. *Cell* **46**: 89.

Lindenbaum, M. and F. Grosveld. 1990. An *in vitro* globin gene switching model based on differentiated embryonic stem cells. *Genes Dev.* **4**: 2075.

Magram, J., K. Chada, and F. Costantini. 1985. Developmental regulation of a cloned adult β-globin gene in transgenic mice. *Nature* **315**: 338.

Mantovani, R., G. Superti-Fuga, J. Gilman, and S. Ottolenghi. 1989. The deletion of the distal CCAAT box region of the Aγ globin gene in black HPFH abolishes the binding of the erythroid specific protein NFE3 and of the CCAAT displacement protein. *Nucleic Acids Res.* **17**: 6681.

Mantovani, R., N. Malgaretti, N. Nicolis, A. Ronchi, B. Giglioni, and S. Ottolenghi. 1988. The effects of HPFH mutations in the human γ globin promoter on binding of ubiquitous and erythroid specific nuclear factors. *Nucleic Acids Res.* **16**: 7783.

Martin, D. and S. Orkin. 1990. Transcriptional activation and DNA binding by the erythroid factor GF-1/NF-E1/Eryf 1. *Genes Dev.* **4**: 1886.

Martin, D., S. Tsai, and S. Orkin. 1989. Increased γ globin expression in a nondeletion HPFH mediated by an erythroid-specific DNA-binding factor. *Nature* **338**: 435.

Mignotte, V., E.F. Eleouet, N. Raich, and P.H. Romeo. 1989a. *Cis-* and *trans-*acting elements involved in the regulation of the erythroid promoter of the human porphobilinogen deaminase gene. *Proc. Natl. Acad. Sci.* **86:** 6548.

Mignotte, V., L. Wall, E. de Boer, F. Grosveld, and P.-H. Romeo. 1989b. Two tissue-specific factors bind the erythroid promoter of the human porphobilinogen deaminase gene. *Nucleic Acids Res.* **17:** 37.

Muller, H., J. Sogo, and W. Schaffner. 1989. An enhancer stimulates transcription in *trans* when attached to the promoter via a protein bridge. *Cell* **58:** 767.

Ney, P.A., B.P. Sorrentino, C.H. Lowrey, and A.W. Nienhuis. 1990. Inducibility of the HS II enhancer depends on binding of an erythroid-specific nuclear protein. *Nucleic Acids Res.* **18:** 6011.

Nathan, D.G. 1990. Pharmacologic manipulation of fetal hemoglobin in the hemoglobinopathies. *Ann. N.Y. Acad. Sci.* **612:** 179.

Pevny, L., M.C. Simon, E. Robertson, W.H. Klein, S.-F. Tsai, V. D'Agati, S.H. Orkin, and F. Costantini. 1991. Erythroid differentiation in chimaeric mice blocked by a targeted mutation in the gene for transcription factor GATA-1. *Nature* **349:** 257.

Philipsen, S., D. Talbot, P. Fraser, and F. Grosveld. 1990. The β-globin dominant control region hypersensitive site 2. *EMBO J.* **9:** 2159.

Poncz, M., P. Henthorn, C. Stoeckert, and S. Surrey. 1989. *Globin gene expression in hereditary persistence of fetal hemoglobin and δβ thalassaemia.* Oxford University Press, United Kingdom.

Pruzina, S., O. Hanscombe, D. Whyatt, F. Grosveld, and S. Philipson. 1991. Hypersensitive site 4 of the human β globin locus control region. *Nucleic Acids Res.* **19:** 1413.

Raich N., T. Enver, B. Nakamoto, B. Josephson, T. Papayannopoulou, and G. Stamatoy-annopoulos. 1990. Autonomous developmental control of human embryonic globin switching in transgenic mice. *Science* **250:** 1147.

Reitman, M., E. Lee, H. Westphal, and G. Felsenfeld. 1990. Site independent expression of the chicken β^A globin gene in transgenic mice. *Nature* **348:** 749.

Romeo, P.H., M.H. Prandini, V. Joulin, V. Vignotte, W. Prenant, W. Valnchenker, G. Marguerie, and G. Uzan. 1990. Megakaryocytic and erythrocytic lineages share specific transcription factors. *Nature* **334:** 447.

Ryan, T.M., R.R. Behringer, N.C. Martin, T.M. Townes, R.D. Palmiter, and R.L. Brinster. 1989. A single erythroid-specific DNase 1 super-hypersensitive site activates high levels of human β-globin gene expression in transgenic mice. *Genes Dev.* **3:** 314.

Schroeder W. 1980. The synthesis and chemical heterogeneity of human fetal hemo-globin. *Hemoglobin* **4:** 431.

Shih, D., R. Wall, and S.G. Shapiro. 1990. Developmentally regulated and erythroid-specific expression of the human embryonic β-globin gene in transgenic mice. *Nucleic Acids Res.* **18:** 5465.

Sorrentino, B.P., P.A. Ney, D.M. Bodine, and A.W. Nienhuis. 1990. A 46 base pair enhancer sequence within the locus activating region is required for induced expression of the γ-globin gene during erythroid differentiation. *Nucleic Acids Res.* **18:** 2721.

Stief, A., D.M. Winter, W.H. Stratling, and A.E. Sippel. 1989. A nuclear DNA attachment element mediates elevated and position independent gene activity. *Nature* **341:** 343.

Talbot, D. and F. Grosveld. 1991. The 5'HS2 of the globin locus control region functions through the interaction of a multimeric complex binding at two functionally distinct NF-E2 binding sites. *EMBO J.* **10:** 1391.

Talbot, D., S. Philipsen, P. Fraser, and F. Grosveld. 1990. Detailed analysis of the site 3 region of the human β-globin dominant control region. *EMBO J.* **9:** 2169.

Talbot, D., P. Collis, M. Antoniou, M. Vidal, F. Grosveld, and D.R. Greaves. 1989. A dominant control region from the human β-globin locus conferring integration site-independent gene expression. *Nature* **338:** 352.

Taramelli, R., D. Kioussis, E. Vanin, K. Bartram, J. Groffen, J. Hurst, and F.G. Grosveld. 1986. γδβ-thalassaemias 1 and 2 are the result of a 100 kbp deletion in the human β-globin cluster. *Nucleic Acids Res.* **14:** 7017.

Townes, T.M. and R.R. Behringer. 1990. Human globin locus activation region (LAR): Role in temporal control. *Trends Genet.* **6:** 219.

Townes, T., J. Lingrel, H. Chen, R. Brinster, and R. Palmiter. 1985. Erythroid-specific expression of human β-globin genes in transgenic mice. *EMBO J.* **4:** 1715.

Tuan, D., W. Solomon, Q. Li, and I. London. 1985. The "β-like globin" gene domain in human erythroid cells. *Proc. Natl. Acad. Sci.* **82:** 6384.

Tuan, D., W. Solomon, I. London, and D.P. Lee. 1989. An erythroid-specific, developmental-stage-independent enhancer far upstream of the human "β-like globin" genes. *Proc. Natl. Acad. Sci.* **86:** 2554.

Wall, L., E. de Boer, and F. Grosveld. 1988. The human β-globin gene 3' enhancer contains multiple binding sites for an erythroid-specific induction of transcription. *Genes Dev.* **2:** 1089.

Wasylyk, B., C. Wasylyk, P. Augerean, and P. Chambon. 1983. The SV40 72 bp repeat preferentially potentiates transcription starting from proximal natural or substitute promoter elements. *Cell* **32:** 503.

Weatherall, D.J. and J.B. Clegg. 1981. *The thalassaemia syndromes.* Blackwell Scientific Publications, Boston.

Whitelaw, E., S. Tsai, P. Hogben, and S. Orkin. 1990. Regulated expression of globin chains and the erythroid transcription factor GATA1 during erythropoiesis in the developing mouse. *Mol. Cell. Biol.* **10:** 6596.

Wright, S., E. de Boer, A. Rosenthal, R.A. Flavell, and F.G. Grosveld. 1984. DNA sequences required for regulated expression of the β-globin genes in murine erythroleukaemia cells. *Philos. Trans. R. Soc. Lond. B* **307:** 271.

Xiao, J., I. Davidson, M. Macchi, R. Rosales, M. Vigneron, A. Staub, and P. Chambon. 1987. In vitro binding of several cell-specific and ubiquitous nuclear proteins to the GT-I motif of the SV40 enhancer. *Genes Dev.* **1:** 794.

Yamamoto, M., L. Ko, M. Leonard, H. Beug, S. Orkin, and J. Engel. 1990. Activity and tissue-specific expression of the transcription factor NF-E1 multigene family. *Genes Dev.* **4:** 1650.

Index

Aminopterin, 6
Androgenetic embryo development, 44
Angelman syndrome, 54–55. *See also* Chromosome imprinting
Animal breeding, 3. *See also* Gene targeting in ES cells
Anterior digit-pattern deformity (*add*) mutation, 30

Backcrosses. *See* Interspecific mouse backcrosses
Bacteriophage λ, 19
Beckwith-Wiedeman syndrome, 56, 64
β-Globin-like gene locus, 100
β-Globin locus regulation, human
  competition model, 105, 107
  developmental regulation, 105–111
  enhancer sequences, 108
  globin gene-switching mechanism, 111–114
  locus control regions, 100–101
  matrix attachment sites, 102
  position independence, 101–105
  protein-binding sites, 103–105
  schematic diagram, 101
  transcriptional activation and/or suppressors, 105
  transgenic mouse models
    β-gene regulation, 110–111
    δ-gene regulation, 108–109
    γ-gene regulation, 109–110
$β_2$-Microglobulin, 7–8
Bovine papillomavirus–metallothionein-growth hormone fusion construct, 30
Buffalo rat liver (BRL)-conditioned medium, 10

Cellular oncogenes, and gene targeting in ES clones, 7
Chromatin
  DNase-I-hypersensitive site in, 19, 26, 102, 112
  loop formation and globin gene switching, 112Chromosome 11, 50, 57, 100
Chromosome imprinting
  definition, 42
  embryonic lethality, 48–49
  genetic techniques to detect imprinting effects, 46
  growth effects, imprinting on chromosome 11, 50
  Igf2 and *Igf2r* genes, 49, 51–52, 56, 62, 64

imprinting and transgenes in mice, 58–59
imprinting in humans
  deletion syndromes, 54–55
  hydatidiform moles, 53
  triploids, 53
  uniparental disomy, 53–54
imprinting regions
  chromosome 2, 47–48
  chromosome 4, 55–56
  chromosome 6, 48, 55
  chromosome 7, 48–50, 56
  chromosome 11, 50, 57
  chromosome 14, 57
  chromosome 17, 50–51, 64
  inactivation, 48–49
  of transgenes in mice, 58–59
  mammalian autosomal imprinting phenomena, 63
  maternal and paternal disomy, 63
  maternal and paternal genomes in mouse development, 43–52
  paternal origin, 46
  role of imprinting, 62–64
  transgenes in mice, 58–59
  tumors, 56–57
  X chromosome, 60–62
  X-inactivation, 61, 64
c-*myb*, 7
Coat color locus in mice. *See* Dilute (*d*) locus in mice
Col1a1 gene, 23, 26
Competition model, developmental regulation of β-globin locus, 105, 107
Copy-number-dependent gene expression, 99, 102–103
Cotransfection in stem cell clones, 5
CpG islands, 82
c-*src*, 7
Cytotoxic T cells, 8

Deletion mapping, 50
Developmental pattern formation in mice, 8
Dilute (*d*) locus in mice, 27
Distonia (*dt*) locus in transgenic mice, 31–32

DNA-binding proteins. *See specific binding protein*
DNA microinjection, 5, 14
DNase-I-hypersensitive sites, 101
  in chromatin, 19, 26, 102, 112
Dominant Acrodysplasia (A*dp*, 30
Downless (*dl*) locus in mice, 30–31
Dystrophia myotonia, 56

Ecotropic murine leukemia virus, 27
Electroporation, 5, 21
Embryonal carcinoma (EC) cells, 21, 74
Embryonic lethality and chromosomal imprinting, 48
Embryonic stem (ES) cells, 14–15, 21, 74, 108
  chromosomal imprinting and, 45
  description, 2
  electroporation of undifferentiated ES cell clones, 21
  gene targeting to null mutations, 4–8. *See also specific oncogene*
  insertional mutagenesis, 20–23
  mouse, 4
  null mutations, 7
  rat vs. mouse, 8–9
  recombinant retrovirus infection, 21
  technical limitations and possible resolutions, 9–10
  transfection and infection, 18
  transgenesis and, 1–8, 21
  X-inactivation and, 91–92
En-2 gene in mice, 8
Endogenous chromosomal genes, 2
Enhancer-trap vectors and insertional mutagenesis in ES cells, 20–22
Erythroid development, 104

Fingerprinting of yeast artificial chromosome (YAC), 90

INDEX 121

Flow-sorting, and X chromosome inactivation, 82–83
Formins, 28. *See also* Limb deformity (*ld*) locus in mice
Fused (*Fu*) gene, 52, 56

G418 resistance in stem cell clones, 5, 21. *See also* neo^r gene
β-Galactosidase gene (*lacZ*), 20
GATA-1, 104–105, 114
Gene targeting in ES cells, 7
 general principles, 5–6
 isolating the targeted clone, 6
 mutagenesis of endogenous genes, 6–8
 screening, 6
Genetic activity and chromosome imprinting, 42
Genetic transformation. *See* Germline transgenesis
Gene-trap vectors and insertional mutagenesis in ES cells, 20–21
 lacZ expression in, 21
Germ-line transgenesis, advantages of tissue-culture system, 2
Globin gene transcription, regulation of, 100–102
β-Globin locus regulation, human
 competition model, 105, 107
 developmental regulation, 105–111
 enhancer sequences, 108
 globin gene-switching mechanism, 111–114
 locus control regions, 100–101
 matrix attachment sites, 102
 position independence, 101–105
 protein-binding sites, 103–105
 schematic diagram, 101
 transcriptional activation and/or suppressors, 105
 transgenic mouse models
  β-gene regulation, 110–111
  δ-gene regulation, 108–109
  γ-gene regulation, 109–110
δ-Globin gene, 99, 105, 108–109
ε-Globin gene, 99–100

γ-Globin gene, 99, 105, 107, 109–110

Hβ58 in transgenic mice, 29
Hairpin tail (T^{hp}) deletion on chromosome 17, 50–51
HD3 breakpoint of mouse X-inactivation center, 77–78, 88
Hemoglobin, 100
Hereditary persistence of fetal hemoglobin, 100, 105–107, 109–110
Heterochromatin, 75
his gene, 23
Histidinol dehydrogenase, 22–23
Homologous recombination, 2
 in ES cells, 18
 gene targeting and, 4–5
Hotfoot (*ho*) locus in transgenic mice, 31
hox-1.5, 8
hprt gene, 28–29, 91–92
Huntington's disease, as candidate for chromosomal imprinting, 55
Hyperplastic egg cylinder (*hec*) phenotype in mice, 32
Hypoxanthine, 6
Hypoxanthine phosphoribosyltransferase (HPRT) function, 6, 28
HPRT deletion mutation, 7

Immune signaling, 7
Imprinting. *See* Chromosome imprinting
Insertional mutagenesis
 examples in mice, 23–32
 generation of, 16–17
 integrity of insertion site, 20
 microinjection of DNA into zygotes, 17, 19–20
 retroviral infection of embryos, 15–18
 retroviral insertional mutagenesis, 19

transgenic techniques, 17–18, 21
Insulin-like growth factor (*Igf2*)
	gene, 49–50, 59, 62
Integration site, 17–18, 20, 104
	of *Mpv17*, 27
	proviral integration site of Mo-MLV, 19
	transgenes in mice, 58–59
International Mouse Genome Mapping Workshop, 84
Interspecific mouse backcrosses, 83–85

Jun/Fos, and human β-globin locus regulation, 103

lacZ gene, 20
Legless (*lgl*) mutation in transgenic mice, 29–30
Lesch-Nyhan syndrome in mice, 28–29
Limb deformity (*ld*) locus in mice, 17, 27–28
Linking clonal libraries and X chromosome inactivation, 82–83, 89
Lipofection, 5
Locus border elements, and human β-globin locus regulation, 102
Loop formation, and globin gene switching, 112

Major histocompatibility complex (MHC) class I genes, 7–8
Mammalian genetics. *See also* Insertional mutagenesis
	advantages vs. disadvantages, 3–4
	gene targeting in ES cells
		general principles, 5–6
		isolating the targeted clone, 6
		mutagenesis of endogenous genes, 6–8
	in mice and insertional mutagenesis, 14–19
Matrix attachment sites, 102

MEL cells, 103
Microdissection and microcloning of X chromosome clones, 80–82
$β_2$-Microglobulin, 7–8
Microinjection, in transgenic mice, 17, 21
Moloney murine leukemia virus (Mo-MLV), 14–15, 21
	proviral integration site, 19
Mouse mammary tumor virus (MMTV)–*myc* fusion constructs, 27–28
Mov13 mutation in mice, 23–26
Mov34 mutation in mice, 26
Mpv17, 26–27
c-*myb*, 7
N-*myc*, 7

neo gene, 21
neo^r gene, 5–6
Neomycin (*neo*) phosphotransferase, 20, 22
Neurofibromatosis, 56
NF-E2, 103–104
N-*myc*, 7
Null phenotype, in ES cells, 7

Oncogenes. *See specific oncogene*
Ornithine transcarbamylase (*Oct*) locus, 75

Parental gametogenesis, 43–45
Parthenogenetic embryo development, 44
Paternal duplication/maternal deficiency and chromosomal imprinting, 49–50
PCR. *See* Polymerase chain reaction
Pedigree analysis, 3
PFGE linkage maps, and X-inactivation center, 89–90
Physical mapping of mouse X-inactivation center region, 89
Plasminogen (*Plg*) gene, 51
Polydactyly, 50

INDEX

Polymerase chain reaction (PCR), 6, 9, 19, 81, 90
Polymorphisms, and animal breeding, 3
Porphobilinogen deaminase (PBGD), 103
Prader-Willi syndrome, 54-55. See also Chromosome imprinting
Pronuclear microinjection, 17-18
Pronuclear transplantation, maternal and paternal genomes in development, 43-45
Proviral integration site of Mo-MLV, 19
Provirus, 15, 18-19, 26
Pulsed-field gel electrophoresis (PFGE), 89
Purkinje cell degeneration (*pcd*) locus in transgenic mice, 31
Pygmy (*pg*) locus in mice, 31

Recessive embryonic lethal mutation, 3
Reciprocal translocations and chromosomal imprinting, 46
Recombination. See Homologous recombination
rec(X) chromosome, 76
Retroviral infection of ES cells, 15-18
Retrovirus, 15
  insertional mutagenesis, 19
Reverse mammalian genetics, 3
Rhabdomyosarcoma, 57

Screening of ES cells, 6
Selection in stem cell clones, 5
Sib selection techniques, isolating the targeted gene, 6
Somatostatin, 31
Spermatogenesis, 30
Spinocerebellar ataxia, 55

c-src, 7
Steinert's disease. See Dystrophia myotonia
Stem cell clones, 5. See also Embryonic stem (ES) cells
Sterility, in mice, 30
STO cells, 10
Superoxide dismutase (*Sod-2*) gene, 51
supF gene, 26

T16H breakpoint, 77, 87-88
T30H breakpoint, 50
Tabby (*Ta*) locus, 78, 85
Targeted mutagenesis of endogenous genes, 6-8
β-Thalassemia, 100, 102-103, 108
Thymidine, 6
Testicular feminization (*Tfm*) locus of X chromosome, 78
Transcriptional activators and/or suppressors of human β-globin locus, 105
Transformation, 5
Transgenesis, 14. See also Insertional mutagenesis
  ES cells and. See Embryonic stem cells
  in mice, 14-15, 21
  model for human β-globin locus, 99-105
Transgenic mice, 1. See also Embryonic stem cells; Transgenesis
Transposon (Tn5), 5
Tumorigenesis and chromosomal imprinting, 56-57

Wilms' tumor, 57
Wnt-1, 7

X-autosome translocations in mice, 75
X chromosome clones, microdissection and microcloning, 80-82

X chromosome imprinting, 60–62
X chromosome inactivation, 61, 74.
See also Chromosome imprinting
 inactivation boundary, 76
 initiation and spreading, 75–77
 mechanisms of X-inactivation, 74–75
 rec(X) chromosome, 76
 translocation breakpoint, 75–76
X-inactivation assay, 90–92
X-inactivation center
 comparative maps, 81
 human, 79–80
 mapping of X-inactivation center, 88–89
 mouse, 77–79, 86
 mapping of X-inactivation center, 84–88
 translocation products, 79

Yeast artificial chromosome (YAC), 90–92
Yolk sac, 44

Zygote, microinjection of DNA, 17–18